Electrocatalysis for Organic Synthesis

Electrocatalysis for Organic Synthesis

DEMETRIOS K. KYRIACOU
DEMETRIOS A. JANNAKOUDAKIS

University of Thessaloniki
Thessaloniki, Greece

A WILEY-INTERSCIENCE PUBLICATION

JOHN WILEY & SONS

New York • Chichester • Brisbane • Toronto • Singapore

Library of Congress Cataloging in Publication Data:

Kyriacou, Demetrios K.
  Electrocatalysis for organic synthesis.

  "A Wiley-Interscience publication."
  Bibliography: p.
  Includes index.
  1. Chemistry, Organic—Synthesis.   2. Catalysis.
3. Electrochemistry.   I. Jannakoudakis, Demetrios A.
II. Title.

QD262.K97 1986   547'.2   85-20239
ISBN 0-471-81247-1

Printed in the United States of America

10 9 8 7 6 5 4 3 2 1

Preface

Electroorganic chemistry has not yet been given its proper recognition in chemical science and technology. This is unfortunate inasmuch as the field has intrinsic value and practical potential. It is, however, encouraging that considerable progress has recently been made toward accepting electroorganic methodology for both laboratory and plant scale synthesis. This new interest in electroorganic chemistry arose from the necessity to confront the increasing scarcity of raw materials and energy and the universal clamor to abate pollution of the environment. It is recognized that although electrochemical methods may be limited by a number of adverse factors, they do possess some very desirable qualities not shared by other methods. Perhaps the greatest practical value of electroorganic chemistry to technology will be sought in the area of electro-catalysis for synthetic purposes.

This book is an introduction to practical electrocatalysis for organic synthesis. As such it is offered primarily to organic synthetic chemists and students who would like to become familiar with recent applications and research goals of modern electroorganic chemistry.

The book consists of three chapters. The first chapter describes the generalized electroorganic reaction and factors affecting it. The second chapter is essentially a working definition of organic electrocatalysis in a broad sense and a discussion of various electrodes and electrocatalysts. The third chapter presents a selected overview of electroorganic and electrocatalytic organic reactions and practical examples illustrating electrocatalytic applications. A list of industrial electrosynthetic processes is included with a comment on the status and future of organic electrochemistry.

DEMETRIOS K. KYRIACOU
DEMETRIOS A. JANNAKOUDAKIS

Thessaloniki, Greece
January 1986

Contents

1. **THE GENERALIZED ELECTROORGANIC REACTION**   1

   1.1 **Heterogeneous Reactions**   2

   1.2 **Homogeneous Reactions**   4

   1.3 **Factors Affecting the Course and Success of Electroorganic Reactions**   4

       1.3.1 Thermodynamic Factors   5

       1.3.2 Kinetic Factors   8

REFERENCES   17

2. ELECTROCATALYSIS FROM AN ORGANIC PERSPECTIVE   **19**

  2.1 Heterogeneous Electrocatalysis   **20**

      2.1.1 Physical Adsorption   **22**

      2.1.2 Chemical Adsorption   **22**

      2.1.3 Electrosorption   **22**

  2.2 Electrodes and Modifications Thereof   **24**

      2.2.1 Physically Activated Electrodes   **24**

      2.2.2 Alloy Electrodes   **26**

      2.2.3 Adatom Electrodes   **27**

      2.2.4 Chemically Modified Electrodes   **27**

      2.2.5 Metal Powder Electrodes   **33**

      2.2.6 Semiconductors as Electrodes   **34**

      2.2.7 Carbon Fiber Electrodes   **35**

  2.3 Homogeneous Electrocatalysis   **35**

  2.4 Electro-initiated Cyclic Reactions; Induced Aromatic Nucleophilic Substitutions   **37**

  2.5 Acid-Base and Ion-Pair Catalytic Effects   **38**

  2.6 Phase Transfer Electrocatalysis   **39**

REFERENCES   **41**

3. OVERVIEW OF ELECTROORGANIC AND ELECTROCATALYTIC REACTIONS OF SYNTHETIC INTEREST   **45**

  A. GENERAL REACTION TYPES   **47**

  3.1 Carbon-Carbon Bond Formation   **47**

  3.2 Carbon-Oxygen Bonds   **52**

3.3   Oxidation of Aromatics, Alkyl
       Aromatics, and Unsaturated Carbon
       Bond Systems   54

3.4   Oxidation of Hydroxy Compounds   57

3.5   Amines and Amides   61

3.6   Anodic Substitutions and Additions   63

3.7   Anodic Fluorinations   66

3.8   Carbon-Hydrogen Bonds and
       Carbon-Nitrogen Bonds   66

3.9   Carbonyl Compounds   68

3.10   Nitro Compounds   69

3.11   Sulfur Compounds   71

3.12   Carbon-Halogen Bonds   72

3.13   Electrocarboxylations   77

3.14   N-Heterocyclic Compounds   80

3.15   Intramolecular Anodic and Cathodic
       Bond Formations   81

3.16   Organometallics   84

B.   SURVEY OF SPECIFIC ELECTROCATALYTIC
       ORGANIC SYNTHESES   86

3.17   Examples of Industrial Electrocatalytic
       Syntheses   86

       3.17.1   Electrohydrodimerization of
                Acrylonitrile   86

       3.17.2   Electrohydrogenolysis of
                Tetrachloropicolinic Acid   87

3.17.3   Indirect Electrosynthesis of
         Calcium Gluconate   89

3.17.4   Indirect Electrosynthesis of
         Dialdehyde Starch   89

3.17.5   Conversion of Propylene to
         Propylene Oxide   90

3.17.6   Oxidation of Toluene to
         Benzaldehyde   91

3.17.7   Electrosynthesis of Vitamin C   91

3.17.8   Oxidation of Anthracene to
         Anthraquinone   92

3.18   Selected Electrocatalyzed Organic
       Reactions   92

3.18.1   Oxidative Electrocatalysis
         (Heterogeneous and
         Homogeneous)   93

3.18.2   Reductive Electrocatalysis   99

3.18.3   Organic Electrocatalysts
         (Mediators)   105

3.18.4   Electrogenerated Bases   109

3.18.5   Superoxide Ion Reactions   111

3.18.6   Chemically Modified
         Electrodes as Electrocatalysts   112

3.19   Solid Polymer Electrolyte Electrolysis   114

C.   INDUSTRIAL STATUS OF ORGANIC
     ELECTROCHEMISTRY IN THE 1980's   116

REFERENCES   121

INDEX   137

Electrocatalysis for Organic Synthesis

CHAPTER 1

The Generalized Electroorganic Reaction

Electroorganic reactions are organic chemical reactions taking place in electrolytic cells by the passage of an electric current. We distinguish two general types of electroorganic reactions: (1) the *direct* or heterogeneous type and (2) the *indirect* or homogeneous type. Figure 1.1 schematically represents these reactions.

1.1 HETEROGENEOUS REACTIONS

A heterogeneous, or direct, electroorganic reaction is one in which the organic substance exchanges electrons directly with the electrode. The fundamental event in such reactions is the electron transfer at the electrode-solution interface. This electrified interface, or electrical double layer, is a very narrow region under the influence of very strong electric fields, as high as 10^7 V/cm. It is the presence of these fields that essentially differentiates heterogeneous electrochemical reactions from heterogeneous thermal or chemical reactions.

An electrodic reaction involves three basic steps:

1. Transfer of electroactive species from the bulk of the solution to the electrode surface or the region of the electrical double layer.

2. Exchange of electrons between electrode and species. Adsorption may be involved here.

3. Removal of the primary electrode products from the electrode surface. Desorption may be involved here.

The slowest of these steps determines the overall reaction rate (current).

Direct anodic reaction

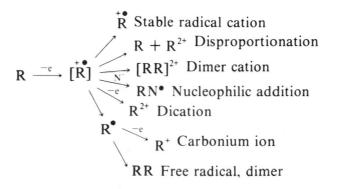

Direct cathodic reaction

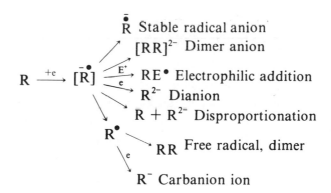

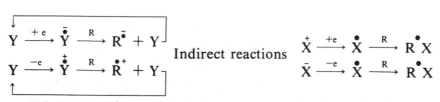

Indirect reactions

FIGURE 1.1 The generalized electroorganic reaction: direct and indirect reactions.

1.2 HOMOGENEOUS REACTIONS

A homogeneous, or indirect, electroorganic reaction is one in which the organic substance does not exchange electrons with the electrode directly but by the intermediation of some electroactive substance. Electron transfers and chemical reactions can then take place in the bulk of the solution until the stable, under the conditions, products are obtained. Indirect reactions also can occur heterogeneously when one or more of the reactants or the electron mediators are adsorbed or immobilized at the electrode surface.

1.3 FACTORS AFFECTING THE COURSE AND SUCCESS OF ORGANIC ELECTROSYNTHESES

Electroorganic reactions are subject to thermodynamic and kinetic factors, as are all kinds of physicochemical processes. Formation of products can be said to be either under thermodynamic or under kinetic control, as depicted in Figure 1.2. If the reactant R affords products R_1 and R_2, the product R_2 with a lower energy level will be thermodynamically more favored than product R_1; but because its activation energy is higher than that for product R_1, it will be formed at a slower rate, and R_1 will be the predominant product. If the reactions are reversible, given sufficient time, the thermodynamically favored product will become predominant. Electroorganic reactions are usually irreversible and therefore their products are under kinetic control in most cases.

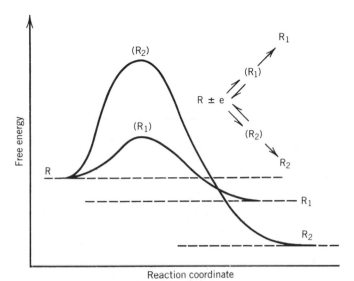

FIGURE 1.2   Formation of products under thermodynamic and kinetic control at a specified electrode potential.

1.3.1   Thermodynamic Factors

1.3.1a   The Electrode Potential

The electrode potential is of fundamental importance in electrochemical reactions since it affects both the energetics and the kinetics of the reactions. Electrosynthetic reactions are driven reactions requiring the imposition of an electric potential in order to occur. The theoretical potential to be applied would ideally be in accordance with the thermodynamic equation relating free energy and potential, that is, $\Delta G = nFE$. An electrochemical reaction would be most desirable if it could be carried out at a potential as close to the thermodynamic potential as possible and at a practical rate, that is, under potentials obeying the fundamental equation

$$E = E^0 - \frac{RT}{nF} \ln Q$$

where the operating potential E would be only a function of the ratio of products to reactants, that is, of $\ln Q$. In practice, however, such a situation is almost impossible. For reversible processes the standard free energy of the reaction is related to the equilibrium constant and to the standard electrode potential for the reaction, that is

$$\Delta G^0 = -RT \ln K = -nFE^0$$

where the symbols have their usual meanings.

In the practice of electrosynthesis, polarographic half-wave potentials, $E_{1/2}$, are valuable concepts and easily obtainable experimental parameters. For practical purposes it can be assumed that

$$nFE^0 = nFE_{1/2}, \quad \Delta E_{1/2} = \Delta\Delta G^0$$

for reversible and quasi reversible processes. Reactions with different ΔG^0 will have correspondingly different $E_{1/2}$. Half-wave potentials of irreversible processes have no thermodynamic meaning, but they are still very useful since they are related to the activation energies of the electroreactions and to their heterogeneous rate constants,

$$\Delta\Delta G^{\neq} = \Delta\ln k = \Delta E_{1/2 \text{ irrev.}}$$

1.3.1b   Reversible and Irreversible Electroorganic Reactions

An electrode process is said to be reversible or not reversible only relative to a time scale used for the experimental observation of the occurrence of the reaction process. For synthetic purposes we would perhaps find it useful to make a distinction between *chemical* and *thermodynamic* reversibility. An electrochemical process may be considered chemically reversible if reversal of the current, at any reachable potential, regenerates the reactants. Thermodynamic reversibility, however, implies regeneration of the reactants in accordance with the Nernstian equation. The two kinds of reversibilities are shown in Figure 1.3.

When the electrode reaction is thermodynamically reversible, the electrode potential is *fixed* by the product/reactant ratio (Q in Nernst's equation). In irreversible processes the potential is not fixed by such equilibrium ratio but is determined by the slowest rate determining step (rds) in the sequence of the steps comprising the entire electrochemical re-

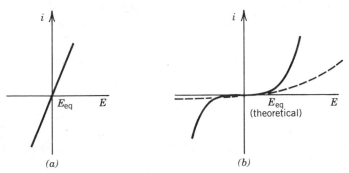

FIGURE 1.3   Thermodynamic (*a*) and chemical (*b*) reversibilities. Dashed line: completely irreversible process.

action. Thus by way of a formal example, the overall electro-chemical reaction process

$$R + 2e + 2H_2O \longrightarrow RH_2 + 2OH^-$$

is irreversible because the two steps following the first reversible step are irreversible.

1. $R + e \underset{rds}{\rightleftharpoons} R^{\cdot -}$ rapid and reversible
2. $R^{\cdot -} + e \xrightarrow{} R^{2-}$ slow and irreversible
3. $R^{2-} + 2H_2O \longrightarrow RH_2 + 2OH^-$ rapid and irreversible

In electrochemical synthesis all steps in the overall reaction must be considered in order to be able to select optimum electrolysis conditions.

1.3.2 Kinetic Factors

1.3.2a Heterogeneous Reaction Rates

As regards the electron transfer itself, it is generally believed that electrons are transferred one at a time, even though the time between successive transfers might be as short as that of a molecular vibration. The rate of an electrode reaction as expressed by the net current is related to the overpotential by the equation

$$\eta = a \pm b \log i$$

known as Tafel's equation, which is a special form of the Buttler-Volmer equation

$$i = i_0\{\exp[\alpha nF\eta/RT] - \exp[(1 - \alpha)nF\eta/RT]\}$$

where i, i_0, and α are observed current, *exchange* current, and transfer coefficient, respectively, and the other symbols have their usual meanings. The overpotential η is the *activation* overpotential, and it is a measure of the change in activation energy for the electron transfer reaction. The actual value of the electrode potential E_a includes the equilibrium potential E_{eq}, the activation overpotential defined as $\eta = E - E_{eq}$, where E is the potential in excess of the E_{eq}, the concentration overpotential E_c, and the ohmic overpotential IR, and any liquid junction potentials which are usually neglected in synthetic work, so that the measured electrode potential in an electrosynthesis is

$$E_a = E_{eq} + \eta + E_c + IR$$

In plotting Tafel lines E_c and IR overpotentials must be absent or minimized. The Tafel equation can be written so as to indicate the theoretical values of the constants a and b as follows

$$\eta = -2.3(RT/\alpha nF) \log i_0 + 2.3(RT/\alpha nF) \log i$$

Because activation overpotentials relative to true equilibrium potentials are very seldom known, the Tafel equation is written as

$$E = a + b \log i$$

where E denotes a relative electrode potential measured against any reference electrode suitable for an electrosynthesis. From such practical Tafel relations it is possible to compare the electrocatalytic properties of electrodes and also to infer plausible reaction mechanisms, and to select optimum applied potentials with regard to rates and selectivities of reactions.

There are some organic reactions that can be carried out very efficiently by simply applying the proper potential. Usually, however, organic molecules are complex species and can react in a number of ways even if the electrode potential is under strict control. In most cases large overpotentials are required to carry out the reaction at a practical rate. Large overpotentials generate *highly energetic primary and intermediate products* and therefore the probabilities for unwanted side reactions are increased. As a rule, the higher the overpotential is, that is, more negative for cathodic reactions or more positive for anodic reactions, the greater the reactivity of the primary products will be, and hence the smaller the reaction selectivity is. Electrocatalysis aims at lowering the overpotential and thus enhancing both the reaction rate and the reaction selectivity. If the electrode is a specific electrocatalyst for a given reaction, setting the electrode potential at a proper value will affect primarily the electronation or deelectronation rate of the desired reaction. Chemical reactions, however, induced by the electrode reaction, cannot be expected to be controlled solely by the potential of the electrode. Other factors must also be considered. Consider, as an example, the symbolic reaction

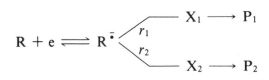

The electrode potential will determine the rate of formation of $R^{\bar{\cdot}}$, its concentration at the electrode-solution interface, and, by virtue of various potential-dependent physical and chemical forces, the reactivity of the primary electrode product with the chemical environment as a whole. The rates of the chemical reactions, r_1, and r_2, will depend on the concentration of $R^{\bullet}$ and also on the concentrations of the species involved in the reactions leading to the intermediates X_1 and X_2 and to the final products P_1 and P_2. If the electrode is a specific electrocatalyst for the desired reaction, for example, for the reaction

$$R + e \rightleftharpoons R^{\bullet -} \longrightarrow X_1 \longrightarrow P_1$$

P_1 will be preferentially obtained. If the electrode is not such a specific catalyst, the rates r_1 and r_2 can be altered relative to each other so as to guide the reaction in favor of the desired end-product. Consideration then should be given to parallel reactions, concurring with, or following, the first electrodic step, with regard to their being heterogeneous or homogeneous or somewhat between both types of reactions. Because the initially produced species are generally very reactive, temperature changes usually exert little effect in changing the rate ratios of the chemical reactions. Concentrations and relative concentration changes can, however, be expected to affect differently chemical reaction rates and thus enhance the overall reaction selectivity. Consideration must be given to the life expectancy of the initial radical ions or radicals. The stability of these species can determine the nature of the chemical reactions. Very short lifetimes will not allow the species to escape into the bulk of the medium and they will be forced to react at the sites where they formed in an adsorbed or unadsorbed state of being. Such chemical reactions will be in effect heterogeneous. Mass transfer effects and current densities, which determine surface concentrations of primary reactive species, would be major controlling factors on the overall reactions. Electric fields and structures of electrical double layers will affect the course of reactions to various degrees in such cases and much more when the life expectancies of the primary electrode products are not long enough as to allow them to escape away from the electrode-solution interface. Life times longer than $\sim 10^{-2}$ seconds might be sufficient for escape into solution; shorter than $\sim 10^{-8}$ seconds may not be sufficient for such escape. Obviously, factors such as stirring (flow rates), temperature, and physical state of the electrolysis medium can variously influence the overall course of the reaction.

CURRENT-POTENTIAL RELATIONSHIP AND REACTION SELECTIVITY

An electrode reaction can be performed either under activation control (Tafel relationship) or under diffusion (mass transfer) control. Substances may have different redox potentials and also different diffusion rates. It would be possible sometimes to improve desired product yield ratios by proper combination of electrode potential and stirring of the solution. It often happens that two reactions at an electrode occur simultaneously but at distinctly different rates. In such cases Tafel lines can be obtained with only one of the reactants at a time. From the two lines thus obtained the optimum potential for the best product ratio can be selected as shown in Figure 1.4. At the electrode potential E_4 reactions (1) and (2) occur at the same rate. If the electrolysis is carried out at that potential the molar ratio B/D would be close to unity, whereas at other potentials various product ratios will be secured.

THE ELECTROLYSIS MEDIUM

The primary electrode products, such as radical ions, ions, or free radicals, could conceivably be formed in any medium.

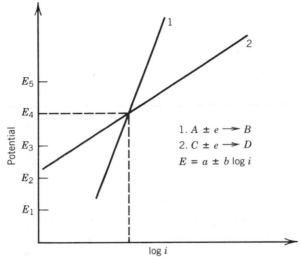

FIGURE 1.4 Tafel lines of simultaneous reactions with different slopes. Crossing point represents equal currents (reaction rates).

However, the chemical reactions that are necessary to generate the desired final products depend on the chemical environment in which the initial electrode species find themselves. This environment consists of the solvent, the supporting electrolyte, and any other organic or inorganic substance that may be needed in the overall scheme of the electrochemical reaction. There are electrochemical reactions in which the kind and concentration of the supporting electrolyte determine the course of the reaction and the nature of the final product(s). Even small amounts of certain impurities or additives can play a determining role in electroorganic synthesis.

The kind and concentration of the supporting electrolyte determine the thickness and compactness of the electrical double layer and the physical and chemical properties of that region of the electrolytic system. In some cases, although the bulk medium may be aqueous, the *hydrophobicity* of the electrode-solution interface may render that part in effect anhydrous and thus allow reactions to occur which could be impossible in an all aqueous medium. Orientation and adsorption forces can also be under the influence of the structure of the double layer. An interesting case arises when large molecules, such as those of natural origin, are to be electrochemically reacted. In such cases if the supporting electrolyte concentration is excessively large, the thickness of the double layer will be considerably decreased so that only a part of the organic large molecule may be within the double-layer zone, the rest being outside of it and therefore affected peculiarly by the prevailing electric field. Such a physical situation may influence the electrochemical reaction favorably or unfavorably. It would be well to note here that the largest electroorganic industrial process for adiponitrile production is practical by virtue of the kind of electrolyte and certain additives employed in the electrolysis medium. The importance therefore of the electrolysis medium should be well recognized and its chemical composition precisely defined for all studies and practical applications in electroorganic chemistry.

1.3.2b   Homogeneous or Indirect Reaction Rates

Homogeneous or indirect electroorganic reactions occur via at least one intermediary heterogeneous step in the overall scheme of the reaction. The heterogeneous reaction may involve an inorganic or an organic species. Obviously, the rate of the overall reaction depends on the rate of the heterogeneous reaction which determines the concentration of the electrolytically formed active agent reacting with the organic substrate. Indirect reactions are in many cases homogeneous electrocatalytic reactions since the directly electroactive species may be recycled and can be used only in catalytic amounts. Indirect synthesis can be achieved by continuously electro-generating one of the reactants: for example, superoxide ion, $O_2^{-\bullet}$, from oxygen bubbled through the electrolysis medium for oxygenations; or by forming NO_2^+ from anodic oxidation of NO_2 for nitrations of aromatics; or as in chlorinations via continuous anodic formation of Cl_2, or $Cl^\bullet$ from Cl^- ion:

$$2Cl^- \xrightarrow{-2e} 2Cl^\bullet \xrightarrow{CH_2=CH-CH=CH_2} ClCH_2CH=CHCH_2Cl$$

Most homogeneous electrocatalyses have been oxidative redox-type mediated reactions. Anodically regenerated oxidants, such as $Cr_2O_7^{2-}$, MnO_4^-, Ce^{4+}, Mn^{3+}, Co^{3+}, $Br^\bullet$, $Cl^\bullet$, I^+, and organic radical cations, for example, (triphenylamine)$^{+\bullet}$, have been used. It would be well to point out that homogeneous electrocatalysis (reductive or oxidative) may give different products, from the same substrate, from what heterogeneous electrocatalysis would do, because the catalytic paths can be different. Metallic oxyanion catalysts can act by oxygen transfer or hydride abstraction from the organic substrate. The indirect oxidation of carboxylates may lead to alkanes and esters rather than to Kolbe dimers, when it is catalyzed by triphenylamine radical cation, mainly because of *relative* concentration effects:

$$\phi_3N: \underset{}{\overset{-e}{\rightleftharpoons}} \phi_3N^{\bullet}_{+} \xrightarrow{RCOO^-} RCOO^\bullet + \phi_3N$$

$$RCOO^\bullet \longrightarrow R^\bullet + CO_2$$

$$RH \xleftarrow{H^\bullet \text{ (solvent)}} \quad \xrightarrow{\phi_3N^{\bullet}_{+}} R^+$$

$$R^+ + RCOO^- \longrightarrow RCOOR$$

References

CHAPTER 1: GENERAL REFERENCES

Baizer, M. M. and Lund, H., Eds., *Organic Electrochemistry*, Marcel Dekker, New York, 1983.

Bard, A. J. and Faulkner, L. R., *Electrochemical Methods. Fundamentals and Applications*, John Wiley & Sons, New York, 1980.

Bockris, J. O'M. and Reddy, A. K. N., *Modern Electrochemistry*, Vols. 1 and 2, Plenum, New York, 1970.

Conway, B. E., *Theory and Principles of Electrode Processes*, Ronald, New York, 1965.

Faulkner, L. R., *J. Chem. Educ.*, **60,** 262 (1983).

Fry, A. J., *Synthetic Organic Electrochemistry*, Harper & Row, New York, 1972.

Kyriacou, D. K., *Basics of Electroorganic Synthesis*, Wiley-Interscience, New York, 1981.

Pletcher, D., *Industrial Electrochemistry*, Chapman and Hall, London, 1982.

Wagenknecht, J. H., *J. Chem. Educ.*, **60,** 271 (1983).

Weinberg, N. L. and Tilak, B. V., *Technique in Electroorganic Synthesis*, Part III, Wiley-Interscience, New York, 1982.

Wend, H., *Angew. Chem.*, **21,** 256 (1982).

Yoshida, K., *Electrooxidation in Organic Chemistry. The Role of Cation Radicals as Synthetic Intermediates*, Wiley-Interscience, New York, 1984.

CHAPTER 2

Electrocatalysis from an Organic Perspective

For the purpose of organic synthesis, electrocatalysis may be defined as the art and science of selectively modifying the *overall rates* of electrochemical reactions so that maximum product yields and total energy savings can be attained. This broad definition is concerned not only with the *pure* electron transfer catalysis but also with all reactions following or concurring with the electron transfer reaction and all the chemical and physical factors as they affect the overall reaction rate and the nature of the final products.

For convenience we distinguish two general types of electroorganic catalysis: the heterogeneous and the homogeneous type.

2.1 HETEROGENEOUS ELECTROCATALYSIS

Heterogeneous electrocatalysis for organic synthesis aims primarily at lowering the overpotential of direct electroorganic reactions. In the broadest sense all electrodes could be looked upon as catalysts insofar as the electrode is only the *site* of the electrochemical reaction and is not consumed in the process. Figure 2.1 shows schematically the various types of heterogeneous catalysis.

Heterogeneous catalysis is assisted by adsorption forces between the organic substrate and the adsorbing electrode phase. Adsorption lowers the activation energy according to how strong the binding is and how it affects the adsorbed species chemically and physically. Electrodes can be regarded as special kinds of catalysts, varying in degree and specificity according to their chemical and physical states of being. In heterogeneous electrocatalysis the goal is to modify or activate the electrode surface so as to facilitate the electroreaction in a selective way.

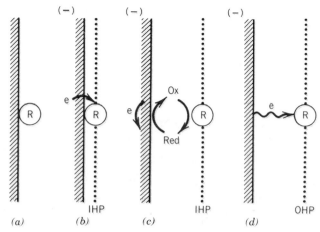

FIGURE 2.1   Pictorial representation of heterogeneous catalysis. (R) is organic reactant. (*a*) Heterogeneous chemical catalysis, adsorption of (R) on catalyst, *no* imposed electric potential, no net charge transfer. (*b*) Heterogeneous electrocatalysis, adsorption of (R) under imposed electric potential, net charge transfer. (*c*) Heterogeneous redox-type electrolysis, redox couple *fixed* at electrode surface, net charge transfer under imposed electric potential. (R) may or may not be adsorbed on electrode (film) surface. (*d*) Direct (unassisted) electron transfer, no adsorption, no catalysis.

A prerequisite for heterogeneous catalysis on *bare* electrode surfaces in general is adsorption of the reacting species onto the electrode surface. For adsorption to be catalytic the transition state between electrode surface sites and reactant(s) would have to be stabilized in order for the activation energy barrier of the surface reaction to be lowered. Strong adsorption usually implies strong catalysis. Adsorption, or electrosorption when it occurs under the influence of an electric field, involves three basic steps: (1) transport of reactants from the bulk of the solution to the surface of the catalyst (electrode), (2) physical and chemical interactions thereupon, and (3) transport of products to the solution. As regards the electron transfer reaction, modern views reject the notion that the catalytic ability of an electrode is a function of the *work function* of the metal and attribute the catalytic properties to the ability of the electrode material to form selectively chemical bonds with the chemical species partaking in the electrochemical reaction.

2.1.1   Physical Adsorption

The causes for physical adsorption are mainly due to electrostatic and lyophobic forces. In addition smaller forces, such as field-dipole interaction and image-potential forces, may be involved.

2.1.2   Chemical Adsorption

Chemical adsorption, also known as activated or specific adsorption, arises through forces similar to those responsible for chemical bond formations. Thus chemical surface compounds can be formed between surface atoms and reactant species, intermediates, and end-products. Although adsorption always plays a role in electrochemical phenomena, it is chemical adsorption via which heterogeneous catalysis occurs. However, even weak physical forces may cause large molecules to orient at the electrode surface in various ways that could enhance or hinder the electron exchange and the selectivity of the reaction.

2.1.3   Electrosorption

Adsorption under the influence of the electric field at the electrode-solution interface is known as *electrosorption*,[1] and it can be schematically shown as

$$\text{electrode} + R_{soln} \rightleftharpoons \text{electrode} \cdots\cdots R_{ads}$$

and under competitive conditions

$$\text{electrode} \cdots\cdots R_{ads} + R' \rightleftharpoons \text{electrode} \cdots\cdots R'_{ads} + R_{soln}$$

Figure 2.2 represents the dependence of adsorption on the electrode potential. As the electrode potential departs from the potential of zero charge (pzc) negatively charged species tend to be adsorbed on the positive side of the pzc while positively charged species on the negative side. Neutral entities will preferentially accumulate near the pzc region. Physical adsorption can affect electrochemical reactions by simply concentrating the reactant species onto the electrode surface.

From what has been considered so far it becomes apparent that adsorption affects the course of electrochemical reactions in two ways: it may accelerate or decelerate reactions, depending on whether adsorption or *desorption* is the rate de-

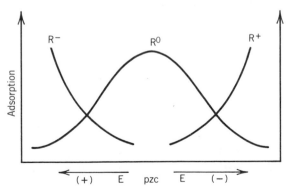

FIGURE 2.2   Electrosorption of charged and uncharged chemical species as a function of electrode potential.

termining steps in the reaction scheme. Adsorption and desorption effects can be observed very conveniently with microelectrodes, for example, polarographically using various electrodes and testing their behavior under conditions similar to those for preparative electrolyses. Although microscale and macroscale results cannot always be related, they can provide preliminary criteria in the search for electrode materials as they do for studying electrolysis conditions and effects of various parameters in general.

2.2   ELECTRODES AND MODIFICATIONS THEREOF

2.2.1   Physically Activated Electrodes

Metallic electrodes exist in various physical states. Nonmetallic electrodes also exist in different states but these states are usually chemical variations of the surface rather than variations in the surface morphology and physical state of the atoms at the electrode's active surface. For synthetic purposes a physically activated electrode is one which by some physical special treatment performs better than the untreated electrode, without apparent chemical changes of its composition. Its catalytic characteristics originate from the physical characteristics of the electrode, namely, crystalline or amorphous states, number and positions of active sites, and nature of the atoms of the metal. Such activations are effected by various methods, for example, polarity reversals, electrodepositions of the same metal, electroreductions of in situ formed surface oxides, photodepositions, and mechanical and chemical treatments of the electrodes.

The overpotential is a function of the current density defined as

$$i = \frac{\text{total current}}{\text{active electrode area}}$$

The physically activated electrode surface displays a much larger area than the inactivated electrode. This results in a decrease of the current density and hence in a decrease of the overpotential for a given total current value or reaction rate. Two different electrodes activated by similar methods may exhibit similar overpotentials for the reduction or oxidation of a polyfunctional molecule but the reaction products may be different. The most practical way to select the most suitable electrode and the method of activation is by actual testing. No activated electrode can preserve its activity indefinitely. Usually the active surfaces are metastable structures. To be practically useful, they should be regeneratable by some simple reactivation method and preferably in situ, or without having to disassemble the cell too often.

Metallic electrodes do not necessarily have to be subjected to any physical activation, as we defined it, in order to be different and more efficient electrocatalysts relative to one another. Such electrodes may still in some way be activated for specific electroreactions by virtue of their nature and their original method of their manufacture. In this sense it is said that platinum is a better electrocatalyst than tin for the evolution of hydrogen from electroreduction of water. Furthermore, platinized platinum and spongy nickel or spongy silver might be better electrocatalysts than platinum and nickel with smooth surfaces. A comment ought to be made regarding overpotential and surface roughness of electrodes:[2] sometimes evolution of gas, for example, H_2 or O_2, may result in over-potentials apparently higher than expected because of trapping of gas bubbles within the irregular crevices of the surface. Reversing the current may reduce this undesirable effect.

2.2.2 Alloy Electrodes

Alloy electrodes have been widely used for inorganic electro-synthesis on an industrial scale. As regards physical activation for such electrodes what has been said for one-metal electrodes applies to alloy electrodes as well. Many types of alloyed materials have been studied in recent years, but not for organic electrochemistry as a goal. Undoubtedly, such electrodes would be found to be specifically useful for certain electroorganic preparations. Some representative examples are: W/C, Cr/Ti, V/Ti, Ni/Fe, mixed oxides, such as $NiCo_2O_4$, $Pb_{0.5}Co_{0.5}CrO_3$, Teflon bonded alloy particles of W/C, Ta/Cr_3C_2, and many other combinations.

In designing alloys to be used as electrodes either in bulk form or as coatings on a substrate material for synthetic reactions, for example

$$A + B \rightarrow AB$$

one considers that the reaction is catalyzed by adsorption of A and/or B and therefore suitable adsorption sites will have to be present. If for example anodic oxygenation of a compound is desired, adsorption sites for oxygen or oxygen affording species will be needed on the electrode surface in order for it to be an active catalyst for the desired oxygenation.

2.2.3 Adatom Electrodes

The underpotential deposition (UPD) of foreign atoms on electrode surfaces has been actively studied in recent years.[3] Electrodes modified by UPD metals display electrocatalytic properties, sometimes accelerating while other times hindering electrode reactions. The electrode surfaces that become covered by *monolayers* or *submonolayers* of UPD metals provide adsorption sites such that the energy of adsorption of the interacting species is altered and the reaction rates are affected depending on the kind of UPD atoms and the extent of surface coverage. In this connection we should bear in mind that foreign adatoms are probably always present on electrodes used for practical electrosyntheses, and it may well be that many electroorganic syntheses reported in the literature have been performed, unknowingly, with UPD modified electrodes. In practice the best we can hope is to be able to reproduce the same *impure* surface, and fortunately this is what usually happens.

The UPD process of metals can be described by a Nernstian-like equation by assuming that the activity of the metal is the activity of a monolayer of metal atoms on a specific metal electrode:[4]

$$E = E^0_{M/M^{n+}} + \frac{RT}{nF} \ln \left(\frac{a_{M^{n+}}}{a_{M_{ads}}} \right)$$

2.2.4 Chemically Modified Electrodes

Chemically modified electrodes represent an entirely new and promising kind of electrodes.[5] As the name implies they have their surfaces modified by attaching to them certain *chemical microstructures* capable of conferring selective and electro-catalytic properties to the electrode. Such electrodes are covered with organic or organometallic films composed of substances that can be attached to the substrate by either chemical or physical bonds. They are also known as *derivatized* electrodes. These electrodes are expected to accomplish the task of homogeneous catalysts but much more efficiently, by having the catalysts immobilized at the electrode surface and in smaller amounts than usually needed for ordinary homogeneous catalysis. Separation of the catalyst from the reaction mixture is greatly simplified since the catalyst remains fixed at the electrode surface. A fundamental prerequisite for a modified electrode is that the film covering it be electrically conducting. So far chemically modified electrodes have not been considered for large scale synthesis because of the instability that has been their main weakness, although there has been some progress in improving their stability. The electrocatalytic activities of these electrodes and the mechanisms therein depend on various factors relevant to the physical and chemical nature of the film. If the films are not permeable to solvent or solvent species, the electrochemical reaction will take place through some intermediation using an organic functional group or a metallic active center incorporated in the film or in the organic molecules that are immobilized at the electrode surface. If the films are more or less permeable, electrochemical reactions may be possible inside the film and also on its surface. Reactions at modified electrodes can be viewed as *outer sphere* or *inner sphere* reactions, for example, for the reaction

Outer
sphere
$$Ox + e \rightleftharpoons Red$$
$$Red + S \xrightarrow{k} Ox + S^- \xrightarrow{k'} product(s)$$

Inner
sphere
$$Red + S \xrightarrow{k} [S \cdot Red] \xrightarrow{k'} Ox + S^- \longrightarrow product(s)$$

where S is the organic molecule. The catalytic system in either mode of reaction provides energy which lowers the overpotential for the electrode reaction. Redox couples in the film act as electron exchange mediators. Biocatalysts, for example, enzymes, are very selective catalysts and have been used to modify electrodes by their incorporation in organic films covering the electrode.

As with homogeneous redox catalysis, catalysis at modified electrodes should be based on kinetic rather than thermodynamic considerations.

2.2.4a Porphyrin Coated Electrodes

Porphyrin coated electrodes are a special kind of chemically modified electrodes. Research in this area is being carried out for the design and synthesis of porphyrin based electrocatalysts aiming at improving the efficiency of economical oxygen cathodes for fuel cell systems. Various metal derivatives of porphyrins, for example, mixed metal "face to face" porphyrin dimers, by analogy to redox enzymes with heterometallic active centers, Co-Ag, Co-Fe, and Co-Mn derivatives, have been synthesized and investigated for their catalytic properties. Extreme variations in catalytic properties have been observed. They have been attributed to the geometry of the compounds, which affects their ability to form the necessary intermediates, such as $Co-O_2-Co$, that are thought to be needed for the four-electron reduction of dioxygen.[6] Thus far most of the work on the potential use of transition metal chelates of N_4 macrocycles as electrocatalysts has been concerned with improving the efficiency of fuel cells. Besides oxygen, other species such as CO, N_2H_4, NO, NH_2OH, alcohols, aldehydes, acrolein, hexacyclodiene, and several others have been found to form adducts with metal phthalocyanines and metal porphyrins.[7] Fe-porphyrins electrocatalyzed alkyl halide reductions.[8]

The aforementioned immobilized redox catalysts (and immobilized enzymes) act by a basic mechanism as formulated below with a macrocyclic iron complex:

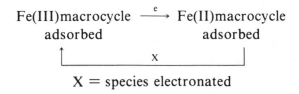

$$\text{Fe(III)macrocycle} \xrightarrow{\;e\;} \text{Fe(II)macrocycle}$$
$$\text{adsorbed} \qquad\qquad\qquad \text{adsorbed}$$

X = species electronated

As it has been mentioned, stability and loss of catalytic activity are still the major unsolved problems before modified electrodes can be used for practical applications.[9] The catalytic activity and stability of redox-modified electrodes and electrodes coated with active polymeric films in general can be observed by cyclic voltammetry, which shows the reversibility of the redox couple attached to the electrode surface, as depicted in the figure. A special type of chemically modified

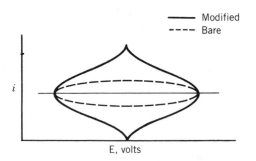

electrodes with synthetic potentialities are the chiral electrodes which are derivatized by attaching to the electrode substrate an asymmetric reagent. Such surfaces are selective toward optical isomer formations.[10]

2.2.4b Electrodes Coated with Polymeric Films

Redox catalysts can be attached to electrodes by incorporating the catalysts to a polymeric film. The films are more or less permeable to electrolyte ions and to solvent molecules. Organic films can be deposited on electrodes from electrolysis media, which may be aqueous, aqueous-organic, or organic. Great attention is given to polymeric films that are electronic conductors, for example, polypyrrole films, which can be prepared in situ by electrochemical polymerization on metal substrates.[11] The polymer itself may contain its own electroactive groups; such a film is one of poly(vinylferrocene). Redox catalysts can be incorporated into the film structure by ion exchange methods. For example, $IrCl_6^{2-}$ can be inserted into poly(4-vinylpyridine) films by ion exchange. The film is first converted into a polyelectrolyte by dipping it into an acidic medium. It is now believed that the electrons are cycled to the redox centers of the films by an *electron hopping* process. Thus the mediated electron transfer from electrode to solution species occurs at a lower potential (electrocatalysis) than direct electron transfer at a bare electrode.

A number of studies have recently been made using the fluoropolymer *Nafion* polymer films.[12] Such films are capable of releasing chemicals (dopants) *on demand* into the surrounding solution. Thus small amounts of ferrocyanide were released by applying pulses of current to a polypyrrole film.[13] Ultrathin organic films have been attached onto electrodes electrochemically. These films are selective toward molecular size and also electric charge on a species.[14] Poly(pyrrole), poly(thiophene) and poly(selenophene) films have been made electrolytically.[15] Direct electropolymerizations have been performed for film preparations using pyrroles, and Co^{2+}, RuO_4^{2-}, Co-porphyrin, and Fe-phthalocyanine as incorporated redox centers in the poly(pyrrole) films.[16]

2.2.4c   Oxide Coated Electrodes

Actually all electrode surfaces exposed to air or to anodic conditions cannot avoid formation of some surface oxides. Probably many anodic oxidations proceed through the intermediation of surface oxides. The successful applications of the *dimentionally stable anodes* (DSA) in inorganic electrochemistry on an industrial scale (chlor-alkali industries) prompted worldwide researches on the preparation and uses of metal supported oxide electrodes; for example, titanium supported RuO_2, IrO_2, and nonstoichiometric oxides RuO_x, $Ir\,O_x$, $RuIrO_x$, and $RuTaO_x$. Several proprietary methods of preparation exist. In the laboratory, oxide electrodes of Ru and Ir can be prepared from their chloride salts by dipping procedures. The electrocatalytic activity of the oxide electrodes was ascribed, among other factors, to the substoichiometric oxidation states of the metal. Oxide electrodes could be in a metastable state, and in practice ways must be available to prolong the active life, or to reactivate the electrodes by some in situ or ex situ method. Potential cycling has in some cases been effective in reactivating the electrodes.[17]

Metal oxide electrodes can become derivatized using silane reagents, as in column chromatography, so as to produce a chemically modified electrode, for example,

Metal/metal oxide couples capable of acting as redox electrocatalysts are the following: $Ti/TiO_2Cr_2O_3$, $Ag/Ag_2O/AgO$, $Cu/Cu_2O/CuO$, $Ti/Mn_2O_3/MnO_2$, $Co/Co(OH)Co_2O_3$, $NiNi(OH)_2Ni(O)OH$, $Pb/PbSO_4/PbO_2$, $Pt/PtO/PtO_2$. Some are stable in acidic and some in basic media.[18]

2.2.5 Metal Powder Electrodes

It is possible to use metal powders as electrodes in the form of slurries or in the form of colloidal suspensions or powders contained in some electrolyte permeable container. The metal powder can be placed in a cloth bag with a *feeder* wire electrode, and the electrolyte passes through the bag. Fluidized bed electrodes operate on this principle. These kinds of electrodes could be recommended when gases are reactants. Colloidal suspensions of semiconductor materials can be used as special forms of electrodes and electrocatalysts in connection with electrophotochemical reactions. Because such microelectrodes expose very large areas, adsorption of reactants is much greater than that with conventional plate or screen electrodes. The ability of colloidal particles to *store* electrons in the solution phase may be found to have interesting applications.[19] Electron transfer from organic radicals to colloidal particles has been shown to be possible. The electrons thus stored on the colloidal particles can be donated to another substance, for example, H_2O, and reduce it to produce another species which may further react homogeneously with some other solution species. Colloidal microelectrodes have been used in the reduction of water, and a theory of their catalytic activity has been proposed.[20]

A special feature of in situ formed metal powder electrodes by electrolytic reduction of the metal ions is that the freshly made particle has a surface free of oxides; such surfaces may act differently from macroelectrodes with not so *clean* surfaces. An interesting study of the activity of freshly generated chromium metal surfaces showed that on the clean electrode surface formed by freshly cut metal, rapid hydrogen evolution occurred from water at potential lower than on the old metal surface.[21] In using powder or colloidal suspension electrodes, the feeder electrode is supplying the electric charge and potential for the suspended catalyst particle as depicted in the figure.[22]

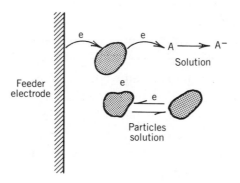

2.2.6   Semiconductors as Electrodes

Semiconductors as electrocatalysts have been used in various forms and ways, and studies thereon have been concerned mostly with saving or converting energy. For electrosynthesis work a semiconductor-electrode combination under proper illumination may be proved valuable in special cases. A photoelectrochemical process at a semiconductor-electrode system can be schematically shown as in the accompanying figure.

2.2.7 Carbon Fiber Electrodes

A potentially useful invention is one concerning the making of carbon fiber electrodes; such electrodes behave like *synthetic metals*, for example, a platinized carbon fiber behaves like a bulk platinum electrode. By *intercalating* various metals, for example, Ag, Cu, and Pb, in graphite fibers, electrodes with catalytic activity have been made for the reduction of nitrocompounds in aqueous media.[23]

2.3 HOMOGENEOUS ELECTROCATALYSIS

In homogeneous electrocatalysis the electrolytic reaction at the electrodes may be only the cyclic electronation and deelectronation of the catalytic substance. Reversible redox sys-

tems, inorganic or organic, can be used as homogeneous catalysts. A homogeneous electrocatalysis can be represented as:

1. Ox + e $\longrightarrow$ Red, at the electrode

2. Red + S $\longrightarrow$ Product + Ox, in solution

3. Ox + e $\longrightarrow$ Red, at the electrode

Catalytic systems such as the following, could be devised:

$$R + e \longrightarrow R^{\overline{\cdot}} \text{ cathode}$$

$$X - e \longrightarrow X^{\overset{+}{\cdot}} \text{ anode}$$

$$R^{\overline{\cdot}} + X^{\overset{+}{\cdot}} \longrightarrow RX \text{ solution}$$

$$RX + AB \longrightarrow Rc + B + X \text{ solution}$$

Substance X is the catalyst continuously regenerated in the
electrolytic reactor. Such systems can be viewed as 200% efficient, as when a reactant is transformed into the same product
in both, the anodic and the cathodic, compartments of the
cell.[24]

Some redox catalytic couples that have been, or are, used
on a large scale are the following:

Br_2/Br^- oxidation of glucose to calcium gluconate

Ce^{4+}/Ce^{3+} oxidation of toluene to benzaldehyde

Co^{3+}/Co^{2+} oxidation of aromatic compounds

Cr^{3+}/Cr^{2+}, Ti^{4+}/Ti^{3+}, Sn^{4+}/Sn^{2+} reduction of nitrocompounds

Cu^{2+}/Cu^+ hydroxylations

IO_4^-/IO_3^- oxidation of starch to dialdehyde starch

The possibility of forming in situ organic electrocatalysts for synthetic purposes has been demonstrated by using aromatic substances like anthracene, benzonitrile, phenanthridine, and other aromatic and heteroaromatic compounds.[25]

2.4 ELECTRO-INITIATED CYCLIC REACTIONS; INDUCED AROMATIC NUCLEOPHILIC SUBSTITUTIONS

Electro-initiated reactions are a unique kind of electrochemical reactions.[26] They may be looked upon as self-sustaining catalytic reactions or as reactions requiring almost zero net current. An example is the phenol synthesis from benzene using the Fe^{3+}/Fe^{2+}-H_2O_2 system.[27]

$$Fe^{3+} + e \longrightarrow Fe^{2+} \quad \text{initiating step}$$

$$Fe^{2+} + H_2O_2 \longrightarrow OH^{\bullet} + Fe^{3+} + OH^-$$

$$HAr + OH^{\bullet} \longrightarrow HArOH^{\bullet}$$

$$HArOH^{\bullet} \xrightarrow{Fe^{3+}} Fe^{2+} + HArOH^+ \xrightarrow{-H^{\bullet}} ArOH$$

Another similar example is the reaction of arenes with hydroxylamine, using the Cu^{2+}-VO^{2+}-V^{3+} system,[28]

$$Cu^{2+} \xrightarrow{e} Cu^+ \quad \text{initiating step}$$

$$Cu^+ + VO^{2+} \xrightarrow{2H^{\bullet}} Cu^{2+} + V^{3+} + H_2O$$

$$V^{3+} + NH_2OH \longrightarrow VO^{2+} + NH_2^{\bullet} + H^+$$

$$HAr + NH_2^{\bullet} + Cu^{2+} \longrightarrow ArNH_2 + Cu^+ + H^+$$

Extensive studies have been done in this area and in connection with nucleophilic substitutions.[29] Nucleophilic substitutions can be achieved by selecting an electrode potential so as to induce the desired substitutions only, while preventing further reduction of the substituted product. If Y is a suitable nucleophile, the catalytic reaction can be formulated as follows:

$$ArX + e \rightleftharpoons ArX^{\bar{\bullet}} \quad \text{initiating step}$$

$$ArX^{\bar{\bullet}} \longrightarrow Ar^{\bullet} + X^- \quad \text{in solution}$$

$$Ar^{\bullet} + Y^- \rightleftharpoons ArY^{\bar{\bullet}} \quad \text{in solution}$$

$$ArY^{\bar{\bullet}} - e \rightleftharpoons ArY \quad \text{electrode}$$

and/or

$$ArY^{\bar{\bullet}} + ArX \rightleftharpoons ArY + ArX^{\bar{\bullet}}$$

2.5 ACID-BASE AND ION-PAIR CATALYTIC EFFECTS

There are cases in which the primary electrode reaction can be facilitated by positive or negative species and by ion-pair formation. Thus polarographic and voltammetric current potential curves shift to less negative or less positive potentials as the pH of the solution changes indicating in a consistent manner the involvement of protons in the rate determining step,

$$E = E_{1/2} + (RT/F) \ln C_{H^+} + (RT/F) \ln \{(i_d - i)/i\}$$

as for instance in the case of the reduction of acetone,

$$CH_3COCH_3 \xrightarrow{H^+} [CH_3COHCH_3]^+$$

$$\downarrow e$$

$$[CH_3COHCH_3]^\bullet \xrightarrow[H^+]{e} CH_3CHOHCH_3$$

The H^+ in this case enters into chemical reaction, but the catalytic effect can be better shown with a metal ion, for example, Zn^{2+}, which only assists the electron transfer,

$$R_2C{=}O \xrightarrow{Zn^{2+}} [R_2C{=}O \cdots\cdots Zn]^{2+}$$

$$\xrightarrow{e} [R_2C{=}O \cdots\cdots Zn]^+ \xleftarrow{e}$$

Ion association with the organic substrate prior to or simultaneously with electron transfer assists in lowering the overpotential and it can be selective in its effect.[30] Thus the reduction of thiophene was facilitated by zinc ions whereas no such effect was noticed in the reduction of chlorobenzene or methylcyclopropyl ketone.[31]

2.6 PHASE-TRANSFER ELECTROLYSIS

Because most organic compounds are not sufficiently soluble in water, two-phase electrolysis for large scale electrosynthesis offers technical and economic advantages.[32] In phase-transfer electrolysis the electro-generated species can be formed either in the aqueous or in the organic phase. The phase-transfer catalysts are usually quaternary-type salts like those used in conventional two-phase organic synthesis. Acetoxylations of

1,4-dimethoxybenzene and other aromatic compounds have been carried out in CH_2Cl_2-H_2O emulsions.[33] Ferrocene was oxidized to ferricenium ion in a two-phase medium as a model system using a *trickle-type* electrode.[34] Electrohalogenation of poly(vinyltoluene) has been attained in two-phase systems, CH_2Cl_2-H_2O and C_6H_6-H_2O. Some variations in the halogenated products were noticed in the two systems.[35]

In phase-transfer electrosynthesis two reactions occur in separate phases: the heterogeneous electrochemical reaction and the homogeneous chemical reaction as depicted below:

1.

$$\text{Ar–CH}_2\text{OH} + \text{OBr}^- \xrightarrow{-\text{H}_2\text{O}} \text{Ar–CHO} + \text{Br}^-$$

Homogeneous reaction
in organic phase

2. $2\text{Br}^- \xrightarrow[\text{anode}]{-2e} \text{Br}_2 \xrightarrow{\text{H}_2\text{O}} \text{OBr}^- + 2\text{H}^+ + \text{Br}^-$

$R_4N^+OBr^-$ transfer catalyst

Heterogeneous electrochemical reaction
in aqueous phase

Organic electrosyntheses in the so-called *ordered systems* can be of practical interest in certain cases. These systems are in a sense similar to two-phase systems; they are micellar solutions, the micelles being cationic, anionic, or nonionic as regards their electrical net charges. The anodic cyanation of dimethoxybenzenes and some other compounds was studied in micellar media. It was observed that cationic micelles favored the cyanation while the other types did not.[36]

References

CHAPTER 2

1. Gileadi, E., Ed., *Electrosorption*, Plenum, New York, 1967; Appleby, A. J., in *Comprehensive Treatise of Electrochemistry*, Conway, B. E. and Bockris, J. O'M., Eds., Plenum, New York, 1983.

2. Kunh, A. T., Yusof, J. B., and Hogan, P., *J. Appl. Electrochem.*, **9**, 765 (1979).

3. Adzic, R. R., Tripkovic, A. V., and Markovic, N. M., *J. Electroanal. Chem.*, **14**, 37 (1980); Kokkinidis, G. and Jannakoudakis, D., *J. Electroanal. Chem.*, **133**, 307 (1982).

4. Schmidt, E., *Helv. Chim. Acta*, **53**, 1 (1970).

5. Murray, R. W., *Accounts Chem. Res.*, **13**, 135 (1980); Miller, L. L. and van der Mark, M. R., *J. Amer. Chem. Soc.*, **100**, 636 (1978); Heller, A., *Accounts Chem. Res.*, **14**, 154 (1981); Bard, A. J., *J. Phys. Chem.*, **86**, 172 (1982).

6. Collman, J. P., Stanford University, private communication; Collman, J. P., Elliot, C. M., Halbert, T. R., and Tourog, B. S., *Proc. Nat. Acad. Sci., U. S. A.*, **74**, 18 (1977).

7. Zagal, J. and Zanastu, S. V., *J. Electrochem. Soc.*, **129**, 2242 (1982); Tezuka, H., Sekiguchi, O., Obkatsu, Y., and Osa, T., *Bull. Chem. Soc.*, Japan, **49**, 2765 (1976).

8. Elliot, C. M., and Marrese, C. A., *J. Electroanal. Chem.*, **119**, 395 (1981).

9. Faulkner, L. R., *Chem. and Engin. News*, Feb. 27, 28 (1984).

10. Miller, L. L. and van der Mark, M. R., *J. Amer. Chem. Soc.*, **100**, 639 (1978).

11. Shih, Y. S. and Jong, M. J., *J. Appl. Electrochem.*, **13**, 23 (1983).

12. Anson, F. C. and Tsou, Y. M., *J. Electrochem. Soc.*, **138**, 113 (1984).

13. Miller, L. L. and Zinger, B., *J. Amer. Chem. Soc.*, **106**, 6861 (1984).

14. William, K. and Murray, R. W., *J. Amer. Chem. Soc.*, **104**, 269 (1982).

15. Diaz, A. F., Kanazawa, K. K., and Gardini, G. P., *J. Chem. Soc. Chem. Commun.*, **1979**, 635; Yoshiro, K., Kaneto, K., Inoue, S., and Tsukagoshi, K., *Japan: J. App. Phys.*, **22**, 1701 (1983); Satoh, M., Kaneto, K., and Yoshiro, K., *J. Chem. Soc. Chem. Commun.*, **1984**, 1627.

16. Bidan, G., Deronzier, A., and Moute, J. C., *J. Chem. Soc. Chem. Commun.*, **1984**, 1185.

17. Mozota, J., Vulkovik, M., and Conway, B. E., *J. Electroanal. Chem.*, **114**, 185 (1980).

18. Fleischmann, M., Korinek, K., and Pletcher, D., *J. Chem. Soc. Perkins Trans. II.*, **1972**, 1396; Beck, F., and Schultz, H., *Electrochim. Acta*, **29**, 1569 (1984).

19. Henglein, A. and Lilie, J., *J. Amer. Chem. Soc.*, **103**, 1059 (1981).

20. Miller, D. S., Bard, A. J., McLendon, G., and Ferguson, J., *J. Amer. Chem. Soc.*, **103**, 5336 (1981).

21. Burnstein, G. T., Kearns, M. A., and Woodward, J., *Nature*, **301**, 692 (1983).

22. van Der Plas, J. F., Barondrecht, E., and Zeilmaker, H., *Electrochim. Acta*, **25**, 1471 (1980).

23. Jannakoudakis, A. D., Theodoridou, E., and Jannakoudakis, D., *Synt. Metals*, **10**, 131 (1984); Theodoridou, E. and Jannakoudakis, D., *Z. Naturforsch.*, **336**, 840 (1981).

24. Chan, R., Jui, H., Ueda, C., and Kuwana, T., *J. Amer. Chem. Soc.*, **105**, 3713 (1983).

25. Lund, H. and Kristersen, L. H., *Acta Chem. Scand.*, Ser. B, **B33**, 495 (1979); Degrand, L. and Lund, H., *C. R. Seances, Acad. Sci. Ser. C.*, **1980**, 291; Doumas-Bouchiat, J. M., M'Halla, F. M., and Saveant, J. M., *J. Amer. Chem. Soc.*, **102**, 3806 (1980).

26. Amatore, C., Pinson, J., Saveant, J. M., and Thiebault, A., *J. Amer. Chem. Soc.*, **103**, 6930 (1981).

27. Steckman, E. and Wellman, J., *Chem. Ber.*, **110**, 356 (1977).

28. Tomat, R. and Rigo, A., *J. Electroanal. Chem.*, **75**, 629 (1977).

29. Andrieux, C. P., Doumas-Bouchiat, J. M., and Saveant, J. M., *J. Electroanal. Chem.*, **123**, 171 (1981).

30. Sasaki, K. and Kunai, A., *Kagaku (Kyoto)*, **38**, 27 (1983).

31. Mairanouskii, S. G. and Kosychenko, L. I., *Sov. Electrochem.*, **16**, 266 (1980).

32. Fleischmann, M., Tennakoon, C. L. K., Banfield, H. A., and William, P. J., *J. Appl. Electrochem.*, **13**, 593 (1983).

33. Ellis, S. R., Pletcher, D., Camlen, P. H., and Healy, K. P., *J. Appl. Electrochem.*, **12**, 693 (1982).

34. Feess, H. and Wend, H., *J. Chem. Techn. Biotechnol.*, **30**, 297 (1980).

35. Matsuda, Y., Morita, M., Yamamoto, H., Watanaba, H., Saita, T., and Akimoto, A., *Denki Kagaku*, **52**, 711 (1984).

36. Laurent, E., Rauniyar, G., and Thomalla, M., *Nouv. J. Chim.*, **6**, 515 (1982).

GENERAL REFERENCES

Anson, F. C., *Proc. Electrochem. Soc.*, **1984**, p84 (review).

Appleby, A. J., in *Comprehensive Treatise of Electrochemistry*, Conway, B. E., Bockris, J. O'M., and Yeager, E., Eds., Plenum, New York, 1983.

Conway, B. E., *Prof. Surf. Sci.*, **16**, 1 (1984) (review).

Pletcher, D., *J. Appl. Electrochem.*, **14**, 403 (1984) (review).

Saveant, J. M., *Accounts of Chem. Res.*, **13**, 323 (1980) (about catalysis by electrodes).

Trassati, S. and Wendt, H., Eds., *Electrocatalysis, Practice Theory and Further Developments*, Pergamon, Oxford, 1984.

Wend, H., *Electrochim. Acta*, **29**, 1513 (1984) (review).

CHAPTER 3

Overview of Electroorganic and Electrocatalytic Reactions of Synthetic Interest

In this chapter we overview general classes of electroorganic reactions of synthetic interest and point out electrocatalytic factors therein. A survey of specific electrocatalytic syntheses is made.

A. GENERAL REACTION TYPES

3.1 CARBON-CARBON BOND FORMATION

Carbon to carbon bond formation is the innermost essence of
organic synthesis. It so happened that electroorganic chemistry began with a typical carbon-carbon synthesis, namely,
the well-known Kolbe synthesis (1849). Faraday first (1834)
observed the evolution of ethane from the electrolysis of aqueous
acetate solutions using platinum electrodes:

$$CH_3COO^- \xrightarrow[Pt]{-e} [CH_3COO]^\bullet_{ads.} \longrightarrow CH^\bullet_{3ads.} + CO_2$$
$$2CH^\bullet_3 \longrightarrow CH_3CH_3$$

This old electrosynthesis is also an excellent example of electroorganic reactions as affected by electrolysis conditions and
electrocatalytic properties of electrodes. The two-step reaction
as shown above occurs best at platinum or irridium electrodes.
Anodes made of other materials fail to give the Kolbe product.
At graphite anodes the radicals formed are rapidly deelectronated further to afford carbonium ions and products thereof,
for example, alcohols, ethers, esters, and olefins. At platinum
anodes in aqueous media the current-potential relationship
(see Fig. 3.1) points out an important fact that kinetics are
more influential than thermodynamics in electroorganic reactions; were it not so, the Kolbe synthesis would not be
possible in aqueous media, because oxygen evolution would
be the reaction most favored thermodynamically. At about
2 V the current density is about 10 mA/cm² in aqueous acetate
solutions, and up to this point the anodic reaction is that of
oxygen evolution. After about 2 V the potential rises suddenly

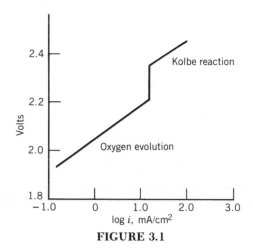

FIGURE 3.1

to about 2.3 V, at which point anodic decarboxylation and formation of the Kolbe product begin. This behavior indicates a drastic change in the *catalytic* ability of the electrode. At lower potentials oxygen evolution occurs, in accord with thermodynamic requirements, while at the more anodic potentials kinetic factors predominate, and the Kolbe reaction occurs at the virtual exclusion of the oxygen evolution.

Oxidation of carboxylates occurs also on graphite anodes, but because graphite adsorbs carboxylate radicals more strongly than platinum does, the radicals are further oxidized to carbonium ions, which then react with the medium to give the oxygenated products. Obviously, extensive adsorption of the free radicals on the electrode surface would enhance both coupling of radicals and carbonium formation, thus yielding a mixture of products.[1]

In explaining the behavior of carbon anodes it has been proposed that carbon surfaces contain paramagnetic centers favoring strong adsorption of the free radicals, which then are in a position to give up another electron to the anode faster than they would on a platinum surface. Thus the anodic decarboxylation of carboxylic acids provides a very good example of how the *catalytic* nature of the surface of an electrode influences the selectivity of the reaction and the nature of the final products.

The anodic condensation of carboxylates, for example, formation of dimethyl sebacate from monomethyl adipate, was influenced by as low as 10^{-4} M concentrations of cyclohexene, styrene, and higher alcohols. Such surface active substances are adsorbed onto the electrode and apparently inhibit the oxidation of solvent and at the same time facilitate the coupling of the radical $R^•$ by weakening the bonds between platinum oxide-radical, PtO-R$^•$, so that the radicals can react more readily with themselves to yield the dimer product[2] RR. Carbon-carbon bonds can form via addition of radicals generated in the Kolbe reaction, thus extending the chain.[3] The Kolbe reaction is indeed a very versatile reaction whereby by incorporating suitable substituents in a carboxylic acid molecule various products can be obtained via decarboxylation.[4]

$$R-\underset{O}{\overset{}{C}}-CH_2CO_2^- \xrightarrow[-CO_2]{-e} R-\underset{O}{\overset{}{C}}-CH_2^• \xrightarrow{X2}$$

$$\longrightarrow R-\underset{O}{\overset{}{C}}-(CH_2)_2-\underset{O}{\overset{}{C}}-R \xrightarrow{H_3O^+} R\underset{O}{\overset{\|}{C}}-(CH_2)_2-\underset{O}{\overset{\|}{C}}R$$

In this connection we note that aromatic carboxylates have not been decarboxylated by the anodic method. Benzoic acid is initially deelectronated to a radical cation that reacts with the solvent,[5] for example, CH_3CN,

The Kolbe synthesis is practiced industrially for the production of diacids and other useful materials. A potentially very important application is receiving much attention recently. It aims at converting biomass to hydrocarbons by first forming carboxylic acids and then decarboxylating anodically. Such processes will require new catalysts, for example, biocatalysts and electrocatalysts along the several stages of the entire process. Carbon-carbon bonds can be formed by other electrochemical reactions as in the coupling of phenols in which both carbon to carbon and carbon to oxygen bonds are formed, the coupling of aryl ethers, the anodic dimerization of certain olefins, and various other electrochemical reactions.[6]

Intramolecular cyclization by anodically forming carbon-carbon bonds in trifluoroacetic acid,—an acid known to stabilize cation radicals, has been demonstrated by the electrolysis of methoxybibenzyls in acetonitrile containing the stabilizing acid.[7] Some anodic intramolecular carbon-carbon bonds have been very efficiently effected, up to 90% yields, as for instance in the formation of dimethoxytetrahydropyrene from metacyclophane.[8]

Cyanations by the anodic method are very effective methods for carbon-carbon and carbon-nitrogen bond formations:

Homolytic substitutions by cyanide are possible but not very likely.

$$CN^- \xrightarrow{-e} CN^\bullet \xrightarrow{RH} RCN + H^\bullet$$

Cyanations may be assisted by the nucleophile itself, as depicted below.

In such reactions the solvent must be a much weaker nucleophile than the CN^- for the yields to be satisfactory, as would
be true for anodic nucleophilic substitutions in general. Cyanations can occur at the aromatic nucleus and on the side chain;
both nitriles and isocyanides can be obtained.[9] Anodic cyanation has been used in the electrochemical synthesis of octa-
alkyl porphyrin.[10]

3.2 CARBON-OXYGEN BONDS

Many classes of compounds undergo anodic carbon-oxygen
linkages via alkoxylations and acyloxylations, where the linkage
can be with the ring or a side chain carbon,[11] for example,

Alcohols anodically oxidized can afford alkoxylation products:

$$CH_3OH \xrightarrow{-e} CH_3O^\bullet + H^+$$

Methoxide ion can be oxidized to the radical also. A meth-
oxylation can thus occur.

$$CH_3O^\bullet + RH \longrightarrow R^\bullet + CH_3OH$$

$$R^\bullet \xrightarrow{-e} R^+ \xrightarrow[-H^+]{CH_3OH} ROCH_3$$

If direct deelectronation of an aromatic molecule occurs at the anode, reaction with methanol or methoxide ion can lead to methoxy products.

$$ArH \xrightarrow{-e} ArH^{+\bullet} \xrightarrow[-e, -H^+]{CH_3O^-} ArOCH_3$$

In such cases the reaction can be catalyzed by the presence of a base, added or generated in situ, and by abstracting a proton from the intermediates.

Anodic acyloxylations can occur in three possible ways.

1. $HAr \xrightarrow{-e} ArH^{+\bullet} \xrightarrow{AcO^-} Ar\overset{\bullet}{H}OAc$

    $\overset{\diagup -e, -H^+}{} $

    $ArOAc$

2. $HAr \xrightarrow{-e} ArH^{+\bullet} \xrightarrow{-e} ArH^{2+}$

    $\overset{\diagup -H^+ +AcO^-}{}$

    $ArOAc$

3. $\left\Vert\, ArH\, \underset{}{|\underline{O}} - \overset{O}{\overset{\|}{C}} - CH_3 \longrightarrow ArOAc \right.$

It may be possible in path 3 that some negative ion that cannot form bonds with the organic substrate will catalyze the de-electronation. Such negatively charged species are always in abundance in the electrical double layer at the anode, and may be expected to assist the electron transfer to the anode. Acetoxylations can occur in media consisting of a single organic solvent or mixtures with a suitable phase transfer catalyst.[12] The electrode potential should be less than 2 V vs saturated calomel electrode (sce) to avoid decarboxylation of acetate.

There are indications that platinum anodes favor acetoxylations, and graphite anodes favor methoxylations.[13] A typical intramolecular acetoxylation is the lactonization of tyrosyl compounds.[14]

It has been observed that surface active agents can affect the selectivity of carbon-oxygen bond formation.[15]

3.3 OXIDATIONS OF AROMATICS, ALKYL AROMATICS, AND UNSATURATED CARBON BOND SYSTEMS

Organic π-systems are good electroactive centers since the electronic structure can be a source or sink of electrons.[1] In the presence of various nucleophiles competitive reactions occur. For example, in the anodic oxidation of anthracene in acetonitrile-pyridine-water media, anthraquinone, bianthrone, and the stable radical cation can be formed.[2]

(in CH_3CN)

In acetonitrile-acetic acid medium with sodium acetate, anodic oxidation of anthracene yields a mixture of *cis* and *trans*,— 9,10-diacetoxy-9,10-dihydroanthracene in a 1:3 ratio of *cis* to *trans* isomers. Oxidation with lead tetraacetate in benzene gives a 1:1 mixture of isomers. This difference can be attributed to electrocatalytic factors associated with adsorption effects in the electrochemical method.

Alkyl substituted aromatic hydrocarbons may lose upon oxidation a proton from the alkyl chain and afford chain substituted products:

$$ArCH_3 \xrightarrow{-e} ArCH_3^{+\cdot} \xrightarrow[-H^+]{-e} ArCH_2^+ \xrightarrow{X} ArCH_2X$$

Thus facile synthesis of aromatic aldehydes can be accomplished,[3] for example,

Unsaturated carbon-carbon molecules with electrodotic groups are anodically oxidized with relative ease. Sometimes a base, added or generated in situ, assists in the oxidation process, for example,

$$
\begin{array}{c}
\underset{NC}{\overset{R_1}{\diagdown}} C = C \overset{R_2}{\underset{OH}{\diagup}}
\quad \rightleftharpoons \quad
\underset{NC}{\overset{R_1}{\diagdown}} CH - C \overset{R_2}{\underset{O}{\diagup}}
\end{array}
$$

$\big\downarrow$ base

$$
\begin{array}{c}
\underset{NC}{\overset{R_1}{\diagdown}} C = C \overset{R_2}{\underset{\underline{O}}{\diagup}}
\quad \overset{-e}{\longrightarrow} \quad
\underset{NC}{\overset{R_1}{\diagdown}} {\bullet}C - C \overset{R_2}{\underset{O}{\diagup}}
\end{array}
$$

$\swarrow$ $-e$

$$
\underset{NC}{\overset{R_1}{\diagdown}} {+}C - C \overset{R_2}{\underset{O}{\diagup}}
\quad \longrightarrow \quad \text{product}
$$

The base acts as a catalyst for the first electron transfer to the anode.[4] The anodic oxidation of aromatics at lead dioxide anodes to produce alcohols and aldehydes is efficiently performed in aqueous acidic media.[5] The oxidation mechanism is usually of the eec or ece type, the slow step being the first electron transfer.

The electrochemical oxidation of allenic hydrocarbons may lead to several products.[6]

Indirect oxidations of alkyl aromatics are apparently very efficient methods. Such electrooxidations have been effected in water emulsions in the presence of transition metal salts, for example, salts of cobalt, cerium, and manganese.[7]

The anodic catalyzed oxidation of o-xylene in CH_3CN-$M(NO_3)_x$-O_2 medium was greatly influenced by the cation of the nitrate electrolyte:[8]

$$NO_3^- \xrightarrow[Pt]{-e} NO_3^\bullet$$

where $M(NO_3)_x = AgNO_3$, $Cu(NO_3)_2 3H_2O$, $Ni(NO_3)_3 6H_2O$, $Fe(NO_3)9H_2O$, and $(n - Bu)_4NNO_3$. The selectivity for o-tolualdehyde was 100% with the quaternary ammonium salt while with the other salts it ranged between 49 and 74%.

3.4 OXIDATION OF HYDROXY COMPOUNDS

Hydroxy compounds are oxidized readily by the anodic method.[1] The anode material and the electrolyte affect the course of the reaction and the nature of the products. Primary alcohols can give acids, aldehydes, acetals, esters, ethers, and hydrocarbons via decarboxylation of carboxylic acids. The oxidation of low molecular alcohols has been widely studied, especially in connection with fuel cells. In such oxidations heterogeneous electrocatalysis on various electrodes has attracted the interest of many workers.[2] Such oxidations proceed via dissociative adsorption and oxidative desorption, that is,

$$R_1CHOHR_2 \longrightarrow R_1\overset{\bullet}{C}OHR_2 + H^+ + e$$
$$\longrightarrow R_1COR_2 + H^+ + e$$

Anodic oxidations of aromatic alcohols in the presence of nucleophiles can become quite interesting synthetic methods. Thus the anodic oxidation of catechol in the presence of 4-hydroxycoumarin, dimedone, and 1,3-indandione gave coumestan derivatives in 90–95% yields using graphite or platinum anodes in water-sodium acetate medium.[3]

The use of nickel salts in the electrolysis medium (10%) has been shown to enhance the oxidation of isobutanol to isobutyric acid in alkaline medium on a graphite anode, giving 90–92% yield of the isobutyric acid. The oxidation in H_2SO_4 medium with PbO_2 anodes is not so selective because of isobutyl ester formation as a side product.[4] Primary and secondary alcohols and α,ω, diols when oxidized on Ni(O)OH anodes yield carboxylic acids, ketones, and dicarboxylic acids (65–85% yields). In hydroxysteroids selective oxidation of 3-OH groups occurs.[5] Secondary alcohols, when subjected to electrolysis in CH_2Cl_2 or $CHCl_3$ in the presence of Ph_2S or Ph_3P and Et_4N tosylate, are transformed to their chlorides, RR′CHCl.[6] The electrocatalytic preparation of pyridine monocarboxylic acids can be achieved on nickel oxide anodes. The oxidation takes place on the anode surface by interaction of the organic molecule with the Ni(O)OH or Ni(III) formed continuously in situ.[7]

The overall electrocatalytic reaction is

Hydroxy compounds with water solubilities of 3 g or less per 100 ml of H_2O can be anodically converted to carboxylic acids on Ni(O)OH electrodes in H_2O-KOH media.[8]

Such typical compounds are chlorophenoxyethanols and chloropyridinoxyethanols, for example,

These anodic oxidations are indirect catalytic oxidations. In acidic media direct oxidations of alcohols are more likely, whereas indirect oxidations are more efficiently performed in basic media. The anodic oxidation of benzyl alcohol to benzaldehyde and benzoic acid can be formulated as follows:

Indirect oxidation of benzyl alcohol by iodonium ion generated from iodide ion at carbon anode in water medium with 2-methyl propan-2-ol as cosolvent affords benzyl benzoate, the yield depending on current density and molar ratio of alcohol to potassium iodide.[9]

Nickel anodes in strong basic media can be effective electrodes for the oxidation of hydroxy compounds. The electrocatalytic Ni(O)OH surface can be easily prepared by electrodissolution of nickel in the alkaline aqueous medium or by precipitation of the colloidal hydroxide on a conducting substrate.[10]

3.5 AMINES AND AMIDES

Aromatic and aliphatic amines are easily oxidized at the anode:

$$\begin{array}{c} H \\ RN: \\ H \end{array} \xrightarrow{-e} RNH_2^{+\cdot} \longrightarrow products$$

Depending on electrolysis conditions a wide variety of products can be obtained.[1] No apparent attempts have been made to electrocatalyze selectively these reactions. Interesting reactions can be formulated, as for example the typical reactions[2] shown on page 62.

The anodic oxidation of aniline in alkaline media gave azobenzene and a polymer with a quinoid structure.[3] Anodic oxidation of aliphatic amines to nitriles at nickel hydroxide electrodes is a very good method, as compared to chemical methods,

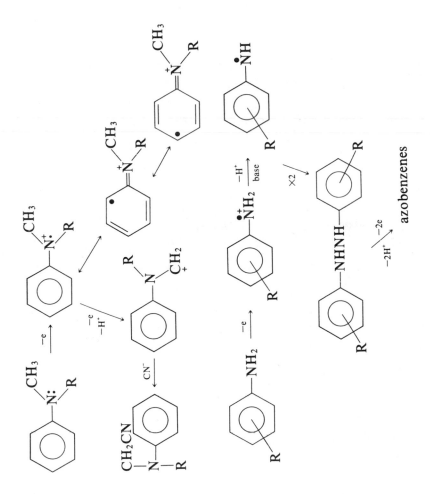

azobenzenes

and may be economically preferable to conventional chemical processes,[4] for example,

85%

Indirect substitutions of aromatic amines can be catalyzed by a halide ion oxidized in situ to halogen[5] or halogen radical. Amides require more positive potentials than amines, primary amides being more difficult to oxidize than secondary and tertiary ones. Anodic oxidation of carbamates prepared from primary amines gave good yields of α-methoxylated compounds,[6] for example, $MeOCHRNHCO_2Me$.

$$R = Bu, n - C_6H_{13}, n\text{-}C_8H_{17}$$

3.6 ANODIC SUBSTITUTIONS AND ADDITIONS

Anodic substitutions can be formulated as

$$RE + N \longrightarrow RN + E^+ + e$$

In most cases the electrophile E^+ is a proton. Nucleophiles are many: H_2O, OH^-, ROH, RO^-, $RCOO^-$, CN^-, NO_3^-, CH_3CN, NO_2^-, OCN^- halide ions, amines, and negatively charged species generated in situ. Substitutions can be direct or indirect:

$$ArCH_3 \xrightarrow[-H^+]{-2e} ArCH_2^{+\bullet} \xrightarrow[-H^+]{H_2O} ArCH_2OH$$

$$Cl^- \xrightarrow{-e} Cl^\bullet \xrightarrow{ArH} ClArH \longrightarrow ArCl + H^\bullet$$

Almost all types of substitutions can be achieved by the anodic method.[1] Typical substitutions resulting in carbon-oxygen and carbon-nitrogen bond formations are the following.

$$ArCH_3 \xrightarrow[-H^+]{-e} Ar\overset{\bullet}{C}H_2 \xrightarrow{-e} Ar\overset{+}{C}H_2 \xrightarrow[-H^+]{CH_3OH} ArCH_2OCH_3$$

$$Ar\overset{+}{C}H_2 + CH_3CN \longrightarrow ArCH_2\overset{+}{N}CCH_3$$

$$\overset{-H^+}{\diagup}\!\!\diagdown_{H_2O}$$

$$ArCH_2NHCOCH_3$$

A case illustrating the effect of the kind of electrolyte on the product distribution is the acetamidation of hexamethylbenzene.

$$Ar(CH_3)_6 \xrightarrow[Bu_4NClO_4]{CH_3CN,\ H_2O}$$

$$Ar(CH_3)_5CH_2OH + Ar(CH_3)_5CH_2NHOCH_3$$

          5%   95%

When tetrafluoroborate ion is used instead of tetrabutylammonium, the product distribution is reversed.[2]

The methodology of phase-transfer catalysis for anodic substitutions is an active field. Acetoxylations and cyanations have been efficiently carried out in emulsions of H_2O/CH_2Cl_2 with a phase-transfer agent.[3] Anodic nitration of 1,4-dimethoxybenzene was performed in micellar medium.[4] Methoxylations have been carried out at a platinum grid anode lying on an ion exchange membrane, thus eliminating the use of a supporting electrolyte. Up to 80% yields were obtained with furan, cyclohexene, and phenylethylene.[5] Intramolecular substitutions have been effected, as for instance in the case leading to cyclization.[6]

3.7 ANODIC FLUORINATIONS

Anodic fluorinations have been well documented.[1] A very recent fluorination study is one concerning the formation of perfluorooctanoyl fluoride and perfluorocyclic ethers from octanoyl chloride.[2] Fluorinations can be performed in acetonitrile containing HF and an amine. Fluoroketones have been obtained by anodic oxidation of enol acetates in CH_3CN-Et_3N-HF medium.[3] At ambient temperatures a yield of 63% PhCOCHFMe from Ph(OAC):CHMe was obtained. Fluorinations take place by nucleophilic attack by fluoride ion on the anodically formed cationic organic substrate. Nickel or monel are the electrodes of choice for electrofluorinations. The fluorinating agent is believed to be the surface NiF_6^{-2} species. Perfluorinated carboxylates, sulfonates, and ethers are made commercially[5] (3M Company, U.S.A.).

3.8 CARBON-HYDROGEN BONDS AND CARBON-NITROGEN BONDS

Many electrolytic organic reductions involve formation of hydrogen bonds. There is no need to discuss such bonds separately. We only indicate the basic types of their formation. Carbon-hydrogen bonds can be formed directly or indirectly,[1] for example,

$$R \xrightarrow{e} R^{\bar{\cdot}} \xrightarrow{H^+} RH^{\bullet} \xrightarrow{e} RH^- \xrightarrow{H^+} RH_2$$

$$Sn^{4+} \overset{2e}{\rightleftharpoons} Sn^{2+}$$

$$Sn^{2+} + RCH_2X \xrightarrow{H_2O} RCH_3 + Sn^{4+} + OH^- + X^-$$

$$\diagdown C = C \diagup \xrightarrow[\text{electrohydrogenation}]{2H^+ + 2e \longrightarrow H_2} \diagdown CH\dot CH\diagup$$

$$\begin{cases} RX + 2e \longrightarrow R^- + X^- \\ R^- + H^+ \longrightarrow RH \end{cases} \text{electrohydrogenolysis}$$

Hydrodimerizations by the reductive method of activated olefins can be valuable synthetic methods. Thus the industrial adiponitrile method has been developed:[2]

$$2CH_2 = CHCN + 2e + 2H^+ \xrightarrow[Q_4N^+]{Cd} NC(CH_2)_4CN$$

This process can be viewed as a special type of electrocatalytic synthesis being selectively possible by electrical double-layer effects in the presence of quaternary ammonium salts.

Intramolecular cyclizations are effected by the cathodic reduction of activated olefins:

$$(CH_2)_n \diagdown ^{CH=CHR_1}_{CH=CHR_2} \xrightarrow[2H^+]{2e} (CH_2)_n \diagdown ^{CH-CH_2R_1}_{CH-CH_2R_2}$$

Some hydrogenations are sometimes occurring by attack of electrogenerated hydrogen atoms or molecules at the electrode surface that acts also as a catalytic surface for the hydrogenation.[3] Important syntheses can be accomplished by electrochemical reduction of carbon-nitrogen bonds. Once a radical ion is formed, various electrophiles can combine with the anion:

$$RC \equiv N - \xrightarrow{e} R\dot CH - \bar N -$$
$$R\dot CH - \bar N - \xrightarrow{E} R\dot CH - NE -$$

Thus pyrrolidine and piperidine derivatives have been prepared from Schiff bases and 1-ω-dibromoalkanes,[4]

$$ArCH{=}NAr \xrightarrow[\text{Br(CH}_2)_n\text{Br}]{\text{2e, }-2\text{Br}} ArCH{-}NAr$$
$$(CH_2)_n$$

3.9 CARBONYL COMPOUNDS

Electrolytic reductions of these compounds can be effected in protic and aprotic media.[1] In protonic media the reduction of ketones can be formulated as an ecc or ecec (e: electrical step; c: chemical step) process:

$$R_1COR_2 + e \longrightarrow R_1\overset{\bullet}{\overset{}{C}}\bar{O}R_2 \xrightarrow{H^+} R_1\overset{\bullet}{C}OHR_2$$

$$R_1CHOHR_2 \xleftarrow{H^+} R_1\overset{\bullet\bullet}{C}OHR_2 \qquad \text{dimer}$$

Aldehydes and ketones are generally more easily reduced than esters and carboxylic carbonyl groups. Such groups can be reduced with relative ease when activated by electron withdrawing groups.

In aprotic media couplings and cyclizations occur.[2]

Cyclic ethers via reduction of ketones in the presence of di-electrophiles can be obtained in one-step syntheses,[3] for example,

$$Ar_2CO \xrightarrow[\text{Br(CH}_2)_n\text{Br}]{2e} \underset{\underset{(CH_2)_n}{|}}{\overset{\overset{Ar}{|}}{Ar-C-O}} + 2Br^-$$

Electroinduced aldol type condensations have been possible by using only catalytic amounts of current, 7.5×10^{-5} F/mol in aldehyde-DMF solutions.

$$RCH_2CHO \xrightarrow[\substack{DMF \\ Pt}]{e} [RCH_2CHO]^{\cdot-}$$
$$\hspace{5cm} EGB$$

The electrogenerated base, EGB, promotes the reaction. α-β-unsaturated carbonyl compounds are thus obtained with very high current efficiencies.[4]

3.10 NITRO COMPOUNDS

The nitro group is one of the best electrophores. A wide variety of reduction products can be obtained by electrolytic reduction of nitro compounds.[1] In anhydrous media nitroalkanes form radical anions that dissociate to yield free radicals and nitrite ions.[2]

$$RNO_2 \xrightarrow{e} R^{\bullet}N\!\!\begin{array}{c} \nearrow O^- \\ \searrow O \end{array} \longrightarrow R^{\bullet} + NO_2^-$$

$$2R^{\bullet} \longrightarrow RR$$

Reduction is possible at all pH's, but the nature of the products depends on both pH and potential:

The effect of agitation of the electrolysis medium has been shown to be of great importance as regards the product distribution.[3] Reduction of nitroolefins affords oximes,[4] for example,

$$PhCH{=}CHNO_2 \xrightarrow[CH_3OH,(H^+)]{e} PhCH_2CH{=}NOH$$

An interesting study of the catalytic effect of Pb and Tl adatoms on the electroreduction of nitro compounds ascribes the catalytic effect to a change of the mechanism, that is from chemical on bare platinum cathode to electron mechanism on adatom-modified platinum (monolayers of lead or thallium atoms).[5]

$$Pt_{bare},\ \phi-NO_2 + 4Pt \longrightarrow \phi-N\underset{*\ \ *}{\overset{O*}{\diagdown}}O_*$$

$$\diagup \overset{2H_{ads}}{-H_2O}$$

$$\phi-NO_* \xrightarrow[H_{ads}]{} product$$

$$Pt/M_{ads},\ \phi-NO_2 \xrightarrow[-H_2O]{2e,\ 2H^+} \phi-NO$$

$$-H_2O \diagdown\ 4e,\ 4H^+$$

$$\phi-NH_2$$

Electrochemical reduction of o-nitroaniline to o-phenylene-diamine was best carried out at a copper cathode as compared to cathodes of Fe, Ni, Pt and Pb, and stainless steel.[6]

3.11 SULFUR COMPOUNDS

Organosulfur compounds can be reduced at various electrodes.[1] Organomercury compounds can be obtained with mercury cathodes. Sulfones yield sulfinic acids and mercaptans. Aromatic sulfones may have their sulfur-carbon bonds cleaved upon electroreduction. Disulfides in aprotic media in the presence of oxygen afford sulfinic acids. Formation of thiolates leads to the sulfinic acids:

$$RS^{\cdot} + O_2^{\cdot -} \longrightarrow RSO_2^-$$

The relative ease of electroreduction of sulfoxides is probably due to the adsorbability of these polar compounds to the electrode surface. Anodic oxidation of organosulfur compounds such as

$$2R_2NC\overset{S}{\overset{\parallel}{-}}S^- \xrightarrow{-2e} R_2NC\overset{S}{\overset{\parallel}{-}}S-S-\overset{S}{\overset{\parallel}{C}}NR_2$$

has been reported to be of industrial interest (Dupont).

Thiolates are often used as protecting groups for carboxylic acids. Deprotection can readily be attained electrooxidatively using bromide salts as electrolytes in H_2O-CH_3CN media,[2] for example,

$$R-\overset{O}{\overset{\parallel}{C}}-S-Bu_t \xrightarrow[Br^-]{-e} RCOOH$$

Oxidative electrodimerization of 1,3-dithiols to tetrathiafulvalenes occurs in CH_3CN media in the presence of pyridine.[3] Selective oxidation of organic sulfides to sulfoxides has been demonstrated by using a polymeric reagent generated and recycled in situ. The catalytic reagent was made from cross-linked poly(4-vinylpyridine).[4]

3.12   CARBON-HALOGEN BONDS

The electroreduction of organic halides and also the halogenation of organic compounds are very active research areas in organic electrochemistry. The general reduction mechanism can be formulated as

$$R-X \longrightarrow \underset{\substack{\text{electrode} \\ \text{(field effect)}}}{R^{\delta+}\cdots\cdots X^{\delta-}} \xrightarrow{e} R^{\bullet} + X^-$$

$$\underset{RR}{\swarrow} \quad \overset{e,\ H^+}{\searrow} RH$$

Evidence of ionic mechanism for the reduction of aromatic halides has been obtained with o-dichlorobenzene in the presence of carbon dioxide.[2]

unstable benzyne
intermediate

In aprotic media reactive intermediates such as benzynes and carbenes can be obtained. Polymeric products are also possible:[3]

(via benzyne)

$$CCl_4 \xrightarrow{2e} \bar{C}Cl_3 + Cl^-$$

$$\bar{C}Cl_3 \longrightarrow Cl\ddot{C}Cl + Cl^-$$

$$Y = CO_2Et$$

Aliphatic halides are reduced more easily, that is, at less negative potentials, than vinyl and aromatic halides. Formation of stereoisomers by the electrolytic method can be expected to be more or less affected by the nature of the electrode and electrolysis conditions:

In aprotic media intermolecular and intramolecular reactions can occur:[4]

$$Br(CH_2)_3Br \xrightarrow{2e} \underset{\underset{\displaystyle CH_2}{|}}{H_2C-CH_2}$$

$$BrCH_2CH_2OH \xrightarrow{2e} \bar{C}H_2CH_2OH + Br^-$$
(I)

$$BrCH_2CH_2OCH_2CH_2OH \xleftarrow[-Br^-]{(I)} \overset{(I)}{BrCH_2CH_2O^-}$$
$$+$$
$$CH_3CH_2OH$$

Very selective electrocatalytic hydrogenolyses can be achieved in certain cases,[5] as in the commercially important case,

$$+ 2Cl^- + 2OH^-$$

$\sim 95\%$

At lead cathodes organolead products are obtained in high yields, for example,

$$4\ CH_3Br + 4e + Pb \longrightarrow Pb(CH_3)_4 + 4Br^-$$

Catalytic effects by a concerted action can be sometimes observed voltammetrically in low dielectric constant media that favor ion pair formation (ion-pair assisted catalysis).

$$\left[>\overset{\cdot}{\bar{C}}Br\cdots\overset{+}{Li} \right] \longrightarrow >C^\bullet + LiBr^-$$

Anodic oxidations of alkyl halides has been explained as follows:[6]

$$RI \xrightarrow[CH_3CN]{-e} [RI]^{\ddagger} \longrightarrow R^+ + I^{\bullet}$$

$$\Big\downarrow \begin{array}{c} CH_3CN \\ H_2O \end{array}$$

$$RNHCOCH_3$$

An interesting selectivity was exhibited in the following case of anodic oxidation in acetonitrile at glassy carbon electrode.[7]

The 1-iodoadamantane (a) gave 80% yield of acetamidated product whereas (b) gave zero yield.

Carbon halogen bonds can also be formed directly or indirectly.[8]

An efficient room temperature chlorination of butadiene to give dichlorobutenes has been achieved in MeCN-CoCl solution as anolyte and ethylene diamine-HCl as catholyte. This method is recommended for large scale application in connection with chloroprene production.[9] In the electrochlorination of o- and p-xylenes the anode materials had an effect on the yield of monochloroxylene, being 66% at platinum, 42–48% at carbon, and 35% at Ti/RuO$_x$ anode.[10]

3.13 ELECTROCARBOXYLATIONS

The electrolytic reduction of carbon dioxide itself is an active field of study.[1] The potential at which carbon dioxide is reduced depends significantly on the catalytic properties of the cathode. On a tin cathode it can be reduced at a potential about 1 V less than on mercury cathode.[2] In aqueous media the reduction can advance all the way to methanol, but in most cases formic and oxalic acids are the major products, the primary reaction occurring as

$$CO_2 + e \longrightarrow CO_2^{\bar{}} \xrightarrow[H_2O]{e} HCO_2^- + OH^-$$

Electrocatalytic reduction of carbon dioxide has been achieved with titanium trichloride, pyrocatechol, sodium molybdate,[3] and with iron-sulfur cluster compounds.[4]

The electrophilic nature of carbon dioxide has been used to produce carboxylic acids by reactions with electrolytically formed organic anions:

$$R{-}X + 2e \xrightarrow{CO_2} RCO_2^- + X^-$$

The carboxylation of α-substituted acetonitriles has been patented.[5]

$$R_1-\underset{\underset{H}{|}}{\overset{\overset{R_2}{|}}{C}}-CN \xrightarrow[Hg]{CO_2} R_1-\underset{\underset{CN}{|}}{\overset{\overset{R_2}{|}}{C}}-COOH$$

Carboxylations of activated carbon-carbon double bonds can be effected using potentials at which only the organic substrate is electronated,[6] for example,

Electroreduction of o-dichlorobenzene in the presence of carbon dioxide yields a mixture of benzoic acid, o-chlorobenzoic acid, and benzene at a mercury cathode in DMF-LiClO$_4$ medium.[7] Also, iodobenzene and bromobenzene yielded benzoic acid when reduced at a mercury cathode in the presence of carbon dioxide.[8]

Electrocarboxylations of unsaturated compounds have been achieved in the presence of methyl chloroformate in acetonitrile medium:[9]

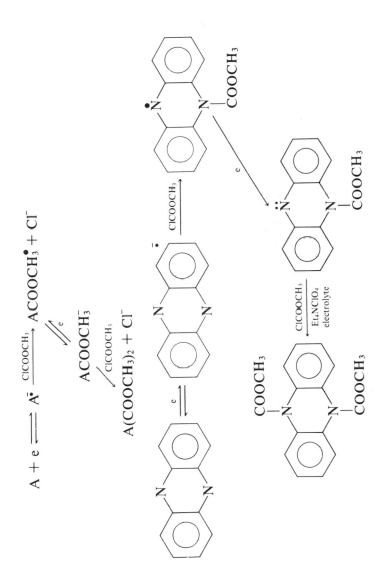

3.14 *N*-HETEROCYCLIC COMPOUNDS

Because of the natural abundance and biochemical importance of *N*-heterocyclics, numerous studies have been made on the electrooxidation and electroreduction of these materials.[1] Their solubility in water makes them favorable candidates for electrochemical studies. In general, the electrochemistry of these molecules is quite complex because of their multifunctionality, for example,

$$R_3 = H, CH_2OH, \text{etc.}$$

The reactions are influenced by the electrode material and the electrolytic medium. In aqueous solution the reduction of 4-aminopyrimidine follows several paths:[2]

Electrochemical reactions of *N*-macrocycles is a very active research area, especially in connection with their applications to chemically modified electrodes.[3]

3.15 INTRAMOLECULAR ANODIC AND CATHODIC BOND FORMATIONS

Intramolecular bond formations are of special interest especially in pharmaceutical synthesis. In many reactions of this kind closely controlled electrode potentials are required. Typical examples are the following:[1]

The ratio of *cis* to *trans* products varies with the kind of acyl groups, supporting electrolyte, solvent, and cathode potential. The cyclization of iodobenzyl isoquinolinium salts is thought to occur as follows:[2]

Because iodine is the easiest leaving halogen, reactions such as that above should be tried via iodo compounds.

Very efficient cathodic intramolecular carbon-nitrogen bond formations are attained with heterocyclic compounds such as the following, in H_2O-sulfolane media and tin cathodes.[3]

85%

In aprotic media electroreductive intramolecular cyclizations via phenyl σ-radicals are possible,[4]

Reactions such as the above would be favored on electrodes that are not very adsorptive for the radicals, so as to minimize further electronation of the radicals.

3.16   ORGANOMETALLICS

Several organometallic syntheses have been reported.[1] The manufacture of tetraethyl lead is well known (as is also the view that this compound should never have been used). The

tetraethyl lead manufacturing process is via the Grignard reagent:

$$4RMgCl + Pb \longrightarrow PbR_4 + 4MgCl^+ + 4e \quad \text{anode}$$

$$4MgCl^+ + 2RCl + 4e \longrightarrow 2RMgCl + 2MgCl_2 \quad \text{cathode}$$

Tetraalkyl derivatives of tin and lead, using alkyl sulfates, were
prepared. Catalytic amounts of alkyl iodides greatly enhance
these reactions. The direct preparation of organometallics using *sacrificial* electrodes has been demonstrated. Such reactions can be explained by a direct metal-radical reaction

$$RX + e \longrightarrow [RX]^{\bar{\cdot}} \longrightarrow R^{\bullet} + X^-$$

$$R^{\bullet} + M_{(electrode)} \longrightarrow MR, \quad \text{e.g.,} \quad C_2H_5Br \xrightarrow[CH_3CN]{Sn} Sn(C_2H_5)_4$$

or by anodic reactions

$$CH_3I \xrightarrow[Sn]{-e} (CH_3)_2SnI_2$$

Very active metallic complexes can be formed directly in the
electrolysis cell for specific organic syntheses. It has been
possible to prepare nickel and iron acetylacetonates in an electrolysis cell in the synthesis of hexadecane from 1-octyl bromide, the complexes playing the role of electrocatalysts, since
in their absence little or no hexadecane was formed.[2] Using a
cadmium anode the β-diketonato complexes of Cd(II) were
obtained in good yields by room temperature electrolysis in
the presence of the desired diketone dissolved in acetonitrile.[3]
Thus complexes of the type Cd(RCOCHCOR')$_2$ were prepared, where R and R' can be CH$_3$, CF$_3$, 2-naphyl, t-C$_4$H$_9$,
and other groups.

B. SURVEY OF SPECIFIC ELECTROCATALYTIC ORGANIC SYNTHESES

This survey is a collection of selected examples that are deemed to be illustrative of the practical merit and potentialities of electrocatalysis for organic synthesis.

3.17 EXAMPLES OF INDUSTRIAL ELECTROCATALYTIC SYNTHESES

3.17.1 Electrohydrodimerization of Acrylonitrile (Monsanto Co.)

In this electrosynthesis the desired reaction is the hydrodimerization of acrylonitrile to adiponitrile.

$$2CH_2=CHCN \xrightarrow[Q_4N^+]{2e} [NCCHCH_2CH_2CHCN]^{2-}$$
$$\swarrow{2H_2O}$$
$$2OH^- + NC(CH_2)_4CN \quad \sim 90\%$$

In the strict sense of the term electrocatalysis, this reaction is not electrocatalytic; but according to our broad definition of organic electrocatalysis, it is a prime example of selective electrosynthesis *catalyzed* by a special modification of the electrical double layer, namely, by Baizer's invention of using a quaternary ammonium salt in the electrolysis medium. The quaternary ion renders the solution layer adjacent to the cathode essentially anhydrous and aprotic, thus minimizing the occurrence of the undesired side reaction,

$$CH_2{=}CHCN \xrightarrow[2H_2O]{2e} CH_3CH_2CN + 2OH^-$$

Catalysis is brought about by water displacement from the electrode region. Such kind of electrocatalysis has been specifically called *electrocatalysis of the second kind*. The cathodes are of bipolar rectangular-type steel plates, coated with cadmium on the negative side, and ~2 mm apart. The concentration of the quaternary salt is a ~0.4%. Apparently this relatively small amount of quaternary salt is sufficient for the desired catalytic selectivity.

3.17.2 Electrohydrogenolysis of Tetrachloropicolinic Acid (Dow Chemical Co.)

This process converts tetrachloropicolinic acid to the 3,6-dichloropicolinic acid with selectivities greater than 90%. The catalytic cathode is a *spongy* silver surface made in situ by polarity changes, positive to negative, in the presence of 10% NaOH in the aqueous electrolysis medium. The key invention in this process is the *in situ formation* of an active, spongy, and silver surface. The reduction can be carried out at several electrodes, but it is either too slow or entirely nonselective. Even smooth or *unactivated* silver cathodes are not electrocatalytic for the desired reaction. The electrolysis occurs in an undivided rectangular cell with parallel electrodes, silver (expanded or screen) as cathode and stainless steel as anode, about 1 cm apart, at 35–40°C. The reactions are as follows:

cathode

$$\xrightarrow[\substack{H_2O \\ NaOH, 10\% \\ -1.3V \ sce}]{4e}$$

$+ \ 2Cl^- + 2OH^-$

anode $4OH^- \xrightarrow{-4e} 2H_2O + O_2 \uparrow$

$+ \ 2OH^- \longrightarrow$

$+ \ 2Cl^- + O_2$

~95%

The electrocatalytic selectivity of this reaction is attributed to specific adsorption forces at the physically activated silver electrode. This electrode, upon activation via periodic polarity changes, takes on a suitable spongy surface onto which oriented adsorption of the organic substrate occurs favoring the formation of the desired dichloro isomer (a very valuable plant growth regulator, trade name Lontrel). The activated silver electrode brings about a lowering of overpotential, as compared with that at mercury cathode, $(E_{1/2} = \sim -1.6 \ V)$, and at the same time it enhances the selectivity of the electro-hydrogenolysis.

3.17.3   Indirect Electrosynthesis of Calcium Gluconate (Sandoz, Switzerland)

The essential fact in this electrooxidation is the use of a bromide ion as a homogeneous mediator,

$$2\ Br^- \longrightarrow Br_2 + 2e \qquad anode$$

which enables the indirect oxidation of glucose to afford very efficiently calcium gluconate.

This industrial process has been in operation since the 1950s.

3.17.4   Indirect Electrosynthesis of Dialdehyde Starch

The electrochemical regeneration of periodate ion to convert starch to dialdehyde starch was seriously considered for large scale production and was used for a period by some industrial concerns but it seems to have lost its momentum for the present (U. S. Department of Agriculture, Miles Laboratories, Hexcel Corp.).

$$IO_3^- \xrightarrow[H_2O]{-2e} IO_4^-$$

Obviously, the thought of using stoichiometric amounts of the oxidizing periodate would be economically unrealistic.

3.17.5   Conversion of Propylene to Propylene Oxide

The conversion of propylene to propylene oxide is another industrial example of indirect electrosynthesis, catalyzed by chloride ion (Bayer-Kellog).

$$2Cl \xrightarrow[anode]{-2e} Cl_2 \xrightarrow{H_2O} HOCl + HCl$$

$$CH_3{-}CH{=}CH_2 \xrightarrow[solution]{HOCl} CH_3CHOHCH_2Cl$$

$$2H_2O + 2e \xrightarrow[cathode]{} 2OH^- + H_2$$

$$CH_3CHOHCH_2Cl + OH^- \longrightarrow CH_3CH{-}CH_2 + H_2O + Cl^-$$
$$\underset{O}{\diagdown\diagup}$$

This indirect electrosynthesis is also a very good example of *paired synthesis* where both electrode reactions yield products needed for the final desired product. Conversion of olefins to their epoxides appears to be more efficiently performed[1] by using Br^- ion as a catalyst instead of Cl^- ion. Propylene oxide can be obtained electrochemically in the presence of bromide ion as a catalyst by using alternating current, as was recently shown.[2]

3.17.6   Oxidation of Toluene to Benzaldehyde

A commercial method using Ce^{4+} as a homogeneous electro-catalyst is the oxidation of toluene to benzaldehyde (Zurich, Switzerland).

3.17.7   Electrosynthesis of Vitamin C

The electrosynthetic method has been successfully used in the manufacture of vitamin C (Hoffman-LaRoche). Diacetone-L-sorbose is oxidized to diacetone-2-keto-L-gluconic acid electrocatalytically on nickel peroxide anode. The catalytic surface peroxide is continuously regenerated in situ by the anodic reaction.

$$Ni(OH)_2 + OH^- \xrightarrow{-e} NiOOH + H_2O$$

The diacetone-2-keto-L-gluconic acid that is produced electrolytically is then converted to L-ascorbic acid (vitamin C).

3.17.8 Oxidation of Anthracene to Anthraquinone

A commercial process using the Cr^{3+}/Cr^{6+} redox catalyst is used for the transformation of anthracene to anthraquinone (Holliday, UK).

Cr^{6+} is regenerated anodically to oxidize homogeneously the anthracene.

3.18 SELECTED ELECTROCATALYZED ORGANIC REACTIONS

For convenience, we discuss several typical examples of electrocatalysis under separate headings.

3.18.1 Oxidative Electrocatalysis (Heterogeneous and Homogeneous)

Heterogeneous anodic electrocatalysis involves, usually, surface oxides of various metals, for example, oxides of nickel, silver, copper, cobalt, manganese, lead, chromium, ruthenium, iridium (DSA), and platinum. Carbon and graphite anodes variously modified and chemically modified electrodes can also be included. Heterogeneous oxidative electrocatalyses have been systematically studied in the case of alcohols and amines using anodes of nickel, silver, cobalt, copper, platinum, and other metals.[3] The oxidation of benzene on a lead dioxide anode in aqueous sulfuric acid emulsions has been done on a semitechnical scale:[4]

$$PbSO_4 + 2H_2O \xrightarrow{-2e} PbO_2 + H_2SO_4 + 2H^+$$

$$3PbO_2 + C_6H_6 + 3H_2SO_4 \longrightarrow 3PbSO_4 + C_6H_4O_2 + 4H_2O$$

$$-2e$$

at anode

Electrocatalysis using metal oxide particles dispersed on glassy carbon surfaces has been observed for the oxidation of ascorbic acid and 1,4-dihydrobenzene. Oxide particles of α-alumina, ZnO_2, Co_2O_3, Cr_2O_3, SnO_2, and ZnO have been used. These oxides give reversible cyclic voltamograms, are easy to prepare, and are very stable.[5] Anodic acetoxylations and cyanations by emulsion phase transfer catalysis may be looked upon as oxidative catalyses. Thus 1,2- and 1,3-dimethoxybenzenes have been cyanated (81% yield) in a $H_2O/CH_2Cl_2/NaCN$ medium. It was observed that wetting factors were important.[6]

It has been established that bromide and iodide ions can be very efficient mediators for the homogeneous oxidation of alcohols and olefinic compounds.[7] The use of the iodide-iodonium redox couple was demonstrated using primary and secondary alcohols:

$$I^- \xrightarrow{-2e} I^+ \text{ at anode}$$

$$RCH_2OH + I^+ \xrightarrow[\text{solution}]{-H^+} RCH_2OI \longrightarrow RCHO + HI$$

$$RCHO + I^+ \longrightarrow R\overset{+}{C}HOI \xrightarrow[-H^+]{RCH_2OH} \underset{\displaystyle \diagdown OCH_2R}{RCHOI}$$

$$\diagup -HI$$

$$RCOOCH_2R$$

$$\underset{R_2}{\overset{R_1}{\diagdown}}CHOH \xrightarrow[-H^+]{I^+} \underset{R_2}{\overset{R_1}{\diagdown}}CHOI \longrightarrow \underset{R_2}{\overset{R_1}{\diagdown}}C{=}O + HI$$

Undivided cells were employed with platinum or graphite electrodes in aqueous media and sodium or potassium iodide as electrolytes and catalysts. The intramolecular electrooxidative cyclization of 2-nitroaniline to benzofuroxan was carried out very readily using iodonium ion formed in situ at graphite anode.[8]

Halide ions have been used as electrocatalysts for the preparation of the sulfoxides of thioformamide derivatives in aqueous media,[9]

Benzyl benzoate was prepared by homogeneous electrocatalysis using I^+ ion as the catalyst:[10]

The dimethyl malonate, DMM, electrohydrodimerization in aqueous hydroxide medium occurs by anodically formed bromine and cathodically formed organic anion[11] of DMM.

$$DMM + e \longrightarrow DMM^-$$

$$DMM^- + Br_2 \longrightarrow DMMBr + Br^-$$

$$DMMBr + DMM^- \longrightarrow DMM\text{-}DMM + Br^-$$

Thus a *paired* electrosynthesis can be achieved using catalytic amounts of bromide ion (a surfactant was also added). Indirect electrochemical oxidation of malonate esters has been performed in nonaqueous solution with iodonium ion as a catalyst:[12]

$$2CH_2X_2 \xrightarrow[I^·]{-e} CHX_2CHX_2 + 2H^+$$

$$X = -COOMe, \qquad -COOEt$$

In this connection it would be well to note that iodide ion can reduce many aromatic cation radicals whereas chloride and bromide ions can do both; they can add and also reduce such radical cations. Thus anthracene and naphthalene can brominate probably by the direct organic cation mechanism by nucleophilic bromide attack on the organic cations.[13] A very practical method of introducing the methoxy group to the α-position of α-amino acid derivatives and α-amino, β-lactams uses Br⁻ or Cl⁻ ions as catalysts.[14] The use of halogen ions as catalysts has been very successful in the formation of N-(cyclohexylthio) phthalimide by electrolyzing a mixture of dicyclohexyl disulfide and phthalimide in acetonitrile medium (99% yield).[15] Furfural was oxidized electrocatalytically by chloride ion in aqueous NaOH-NaCl medium (55–60%).[16]

Several metallic ions that are known to be good oxidants have been employed as homogeneous electrocatalysts. Ceric ion was used in the indirect oxidation of D-gluconic acid and derivatives of D-arabinose.[17] Also the indirect oxidation of δ-D-glucolactone to D-arabinose in aqueous sulfuric acid in the presence of catalytic amounts of ceric ion was found to be a very efficient reaction.[18] Aqueous sulfuric acid media are good for a number of electrocatalytic oxidations of organic materials. Thus aniline was oxidized to benzoquinone in the presence of manganese ions in 20% H_2SO_4 solution.[19] Terephthalic acid was obtained from catalytic oxidation of p-xylene in H_2SO_4-

CrO_3 medium in which the CrO_3 catalyst was formed electro-lytically from $Cr_2(SO_4)_3$.[20] The electrocatalytic preparation of saccharin from o-toluenesulfonamide via chromic ion in H_2SO_4 media has been patented. The chromic ion is regenerated at the anode.[21] Anisaldehyde was prepared electrocatalytically from oxidation of p-methoxytoluene in H_2SO_4-H_2O-Mn^{2+} me-dium.[22] The conversion of benzyl alcohol to benzaldehyde and of anthracene to anthraquinone in good yields has been carried out in sulfuric acid media in the presence of Ag^+ and Cr^{3+} ions. The Ag^+ ion is a good catalyst for the oxidation of Cr^{3+} to dichromate ion at the platinum anode. The dichromate ion is the regenerated catalyst for these oxidations.[23] Electrogen-erated cuprous ion is a strong oxidizing agent for aromatics in the presence of oxygen. Benzene in the presence of cuprous ion and oxygen in H_2O/H_2SO_4/acetonitrile medium was thus oxidized to yield phenol as the major product.[24] Selective elec-trochemical oxidation of alcohols has been possible by using osmium(IV) catalysts of the type bis(pyridine) N, N'-bis(3,5-dichloro-2-hydrobenzamide-1,2-ethane osmium(IV). Thus benzyl alcohol in the presence of the catalyst afforded selec-tively benzaldehyde, with more than 40 catalyst turnovers.[25] Catalytic oxidation of ascorbate[26] by $Os(bpy)_3^{2+}$, (bpy = 2,2'-bipyridine), incorporated on *Nafion* coated electrode was at-tributed to the mediation of the immobilized couple Os^{2+}/Os^{3+} by applying a potential suitable for the reversible re-action $Os^{2+} \longrightarrow Os^{3+} + e$. A very interesting electrocatalytic oxygenation of several hydrocarbons has been accomplished with a manganese porphyrin/periodate system under condi-tions of phase transfer catalysis. Epoxides, up to 90% yields, were obtained. Unactivated alkanes were hydroxylated, giving yields between 25 and 77%. The periodate was regenerated electrolytically in the aqueous phase.[27] This system was said to possess potentialities for large scale applications. Generation of catalytic species for olefin metathesis has been shown to be feasible using WCl_6 or $MoCl_5$ for the in situ production of the active species in CH_2Cl_2 as solvent. In this solvent at an aluminum anode metallocarbene intermediates are formed, which then enter into the overall reaction scheme as shown below with 2-pentene as the olefin:[28]

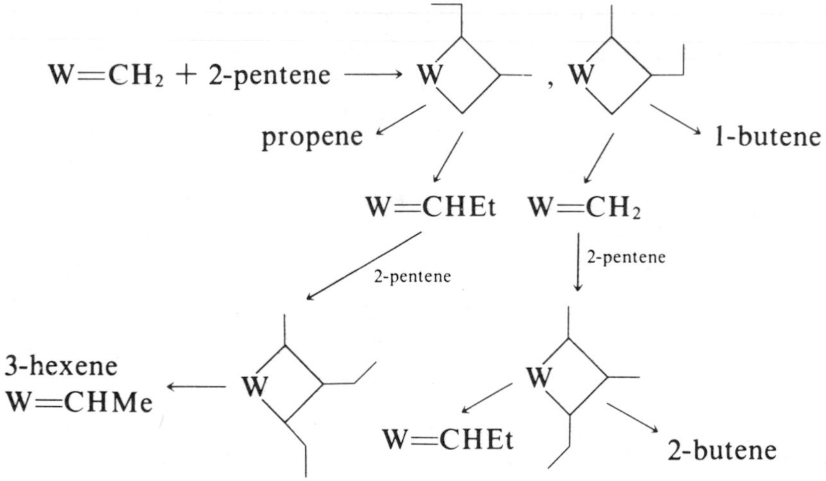

At the aluminum anode the metallocarbene is regenerated via several chemical and electrochemical steps. Olefins can be transformed to allylic derivatives electrocatalytically using as catalysts selenenylating reagents and by an oxyselenenylation-deselenenylation process. Only catalytic amounts of a diphenyl diselenide are required.[29] Methanol or CH_3CN-H_2O solvents are used. Oxyselenenylation occurs via the electrogenerated reagent PhSeOH or PhSeOMe. In methanolic media with $CuCl_2$ as catalyst, electrocatalytic desulfurization of α-(2-benzothiazolylthio) alkanoates to give α,α-dimethoxyalkanoates was possible.[30] Ruthenium type catalysts have been used as homogeneous electrocatalysts. Thus electrocatalytic oxidations of alcohols, aldehydes, and cyclic ketones were carried out with RuIV complexes regenerated anodically in aqueous and in dimethylsulfone-water media.[31] A Japanese patent describes the use of electrolytically regenerated higher valence ruthenium compound for the mild oxidation of α-L-sorbose to sugar lactone or sugar carboxylic acid. Carboxylic acids from alcohols are also made by this method.[32]

3.18.2 Reductive Electrocatalysis

Electrocatalytic heterogeneous cathodic reactions have been more popular than anodic ones perhaps because of the greater availability of cathode materials than materials for anodes. Thus cathodic reactions can be carried out on many electrodes— almost all kinds of metals and their alloys can be used, in order to select the most desirable electrode for the reaction. For example, in the case of the electrochemical reduction of acetone, the products can vary depending on electrode material:

$$
CH_3COCH_3
\begin{cases}
\xrightarrow{\text{Hg}} (CH_3)_2CH\text{—}Hg\text{—}CH(CH_3)_2 \\[6pt]
\xrightarrow{\text{C,Hg,Pb}} CH_3\underset{\underset{OH}{|}}{\overset{\overset{CH_3}{|}}{C}}\text{—}\underset{\underset{OH}{|}}{\overset{\overset{CH_3}{|}}{C}}\text{—}CH_3 \\[6pt]
\xrightarrow{\text{Cd}} CH_3CH_2CH_3 \\[6pt]
\xrightarrow[\text{via ads.,H}^\bullet]{\text{Pt,Ni}} CH_3\overset{\overset{H}{|}}{C}OHCH_3
\end{cases}
$$

Heterogeneous reductions mediated by adsorbed hydrogen on the electrode surface are catalytic reactions, the degree of catalysis depending on the electrode material and its physical and chemical state. Direct and indirect hydrogenations occurring on active surfaces, such as those of black platinum and Raney-metal, are typical heterogeneous electrocatalytic reductions. At platinized platinum, for instance, coadsorption of reactants (electrosorption) can be formulated as

$$R_{ads} + H^{\cdot}_{ads} \rightleftharpoons RH^{\cdot}_{ads} \longrightarrow products$$

Ethylene is thus reduced at platinized platinum at only -0.1 V vs normal hydrogen electrode (nhe), whereas a potential of about 2 V is required at a mercury cathode. The organic substance may dissociate upon adsorption (dissociative chemisorption); and the degree of dissociation, and hence catalysis, depends on the *activity* of the electrode surface. The rates of reactions would depend on the concentration of adsorbed species, so that the rate or current in a hydrogenation reaction can be given by an equation such as

$$rate \propto i = k\ \theta(H^{\cdot}_{ads})(R^{\cdot}_{ads})\theta$$

where $R^{\bullet}_{ads}$ is a species resulting from the dissociation of the adsorbed organic molecule. Electrocatalytic reduction of arylketones on platinum black can proceed all the way to give saturated hydrocarbons.[33]

$$R_1R_2C{=}O \xrightarrow[\text{black}]{\text{Pt}} R_1R_2CH_2$$

The ease of molecular hydrogen addition to unsaturated compounds depends on the catalytic activity of the electrode on which hydrogen is generated. Thus electrodeposition of nickel on Al, Cu, Ni, and Sn on nickel in DMF solvent indicated that the greatest catalytic activity was exhibited on the nickel-copper electrode.[34] The electrohydrogenolysis of certain polychlorinated pyridine compounds on activated (spongy) silver cathodes are prime examples of heterogeneous electrocatalysis: for example, the selective electrohydrogenolysis of tetrachloropicolinic acid to the 3,6-dichloro isomer. Pentachloropyridine reduced at mercury or lead cathodes gives exclusively symmetrical tetrachloropyridine, whereas reduction at spongy silver affords a 1:1 mixture of symmetrical tetrachloropyridine and 2,3,5-trichloropyridine.[35] The adsorptive qualities of the electrodes in connection with the prevailing electric fields at the electrode-solution interface can be very influential in the reaction course and the nature of the products. Electrodes immersed in electric fields provide sites in which molecular *distortions* and even ionizations of certain polarizable bonds can occur, as the following actual case illustrates.[36]

$$BrCH_2(CH_2)_2CHBrCO_2Et$$

$$\downarrow 2e,$$

$$BrCH_2(CH_2)_2\overset{(-)}{C}HBrCO_2Et + Br^-$$

$$
\begin{array}{ll}
\begin{array}{l}
H\diagdown \\
H\!-\!\!\underset{/}{C}\!-\!Br \\
(CH_2)_2\diagdown \\
\qquad\qquad \overset{(-)}{C}H\!-\!CO_2Et
\end{array}
&
\begin{array}{l}
H\diagdown \\
H\!-\!\!\underset{/}{C}\underset{\delta+}{\cdots}\underset{\delta-}{\cdots}\overset{}{Br} \\
(CH_2)_2\diagdown \\
\qquad\qquad \overset{(-)}{C}H\!-\!CO_2Et
\end{array} \\
\quad (a)\ E = -1.5\ V & \quad (b)\ E = -2.5\ V
\end{array}
$$

It was proposed that the reaction leading to the monobromo compound from the dibromobutyrate compound is assisted by the more negative potential at the electrode, as is pictorially shown in scheme *b*, by the field-induced ionization tendency of the C-Br bond. This effect favors the cyclization reaction more than the hydrogenation, which occurs preferentially at the less negative potential, as shown in scheme *a*. The electrode potential can influence the relative differences observed in electrochemical reactions capable of giving *cis* and *trans* isomers, as was shown with the meso and DL-isomers 3,4-dibromohexane, when it was reduced in DMF and NH_3 media.[37]

The electrode potential jointly with adsorption effects may have sometimes surprising effects on the nature of the products.

It was shown recently that Birch type reductions can be performed electrochemically in very basic media with tetraalkylammonium electrolytes.[38] Benzene and steroids were thus reduced to dihydroaromatic products. Tetraalkylammonium amalgams are formed via which the reductions occur. Benzene was thus reduced in an H_2O-DMF-THF mixed medium to give 1,4-dihydrobenzene. The steroid estrone 3-methyl ether was reduced as shown,

and dihydrotoluene was obtained from benzylmethyl ether,

The cathodic pinacolization of carbonyls by electrogenerated Cr(II) and Cr(III) has been shown to occur via complexations, as depicted below.[39]

$$Cr(III) \xrightleftharpoons{e} Cr(II)$$

$$\downarrow nR_2CO$$

$$Cr(II)(R_2CO)_n \xrightarrow{Cr(III)} [Cr(III)(R_2CO)_n] + Cr(II)$$

$$\swarrow e$$

$$[Cr(III)(R_2CO)_n]^{\overline{\cdot}}$$

$$\swarrow^{n=1} \qquad \searrow^{n=2}$$

$$R_2CO^{\overline{\cdot}} + Cr(III) \qquad\qquad Cr(III) \diagup\begin{matrix} O-CR_2 \\ | \\ O-CR_2 \end{matrix}$$

$$\searrow H^+ \qquad \swarrow 2H\cdot$$

$$R_2C(OH)C(OH)R_2$$

Transition metal complexes have been claimed to enhance the electrochemical dicarboxylation of organic compounds[40] in the presence of CO_2. Dicarboxylic acids from unsaturated compounds, for example, butadiene, were thus obtained by electrolyzing at potentials lower than those for CO_2 electroreduction. Mercury cathodes and the complex Fe-dicyclopentadienyl tetracarbonyl were used in tetrahydrofuran with tetrabutyl quaternary salts to convert butadiene to 3-hexene-1,6-dicarboxylate.

A great deal of work has been done with the electrochemical reduction of nitrocompounds and several large scale applications have come into existence over the years. Electrocatalytic possibilities have not apparently been exploited yet in this area. A recent Indian patent claims the use of Sn(II) and Th(I) salts as electrocatalysts for the reduction of nitrobenzene to p-aminophenol in suspension of nitrobenzene in mineral acids.[41] 4,4'-Dinitrostilbene-2,2'-disulfonic acid was converted to the diamino compound, using $Ti_2(SO_4)_3$ electrogenerated from $Ti(SO_4)_2$. Yields up to 90% and current efficiencies of 85% were achieved.[42] Alkenes and aromatic halides can couple electrocatalytically through the assistance of nickel complexes, $(NiI_2(PPh_3)_2)$ acting as catalysts. Thus ethylene and iodobenzene in tetrahydrofuran afforded styrene in 80% yields.[43] Using ethylene and aryl halides, the synthesis of 1,1-diarylethane was catalyzed by $NiBr_2$ in THF/HMPA medium.[44] The catalytic species seems to form by electroreduction of $NiBr_2$. Diarylethane was thus obtained in 65% yield from bromobenzene and ethylene in the presence of Bu_4NBr as electrolyte, the electrolysis being carried out at 0–5°C.

In aprotic media the reduction of aromatic ketones is electrocatalyzed by $MnCl_2$. The Mn(I) is believed to be the catalytic species which is formed in situ and complexed with the ketone. Formation of Mn(O)-ketone complexes and colloidal particles of electroreduced manganese ions adhering at the electrode surface were also thought to be alternative possibilities for the observed catalysis.[45]

3.18.3 Organic Electrocatalysts (Mediators)

Organic mediators with catalytic activities have been effective in a number of electroorganic syntheses.[46] For example, N-hydroxyphthalimide is an effective catalyst for the homogeneous oxidation of hydroxy compounds to aldehydes and ketones. Thus electrolysis in acetonitrile containing the catalyst and pyridine transformed $EtCH_2MeOH$ to EtCOMe in 85% yield. Organic catalysts can be added or electrogenerated in

situ for a specific reaction. They may be fairly stable species
or have only transitory existence, and as such they may not
be available commercially or stored in the ordinary chemical
store room. To function as catalysts they must be capable of
continuous regeneration in the electrolysis reactor; and as redox couples they must be reversible. Several organic substances have been recognized as reversible redox couples with
catalytic properties. Anthracene, phenanthridine, and benzonitrile, are such substances. These substances have been shown
to catalyze the reduction of halopyridines and halobenzenes
by a mechanism thought to be as follows:

$$Ar + e \rightleftharpoons Ar^{\bar{\cdot}} \text{ catalyst}$$
$$Ar^{\bar{\cdot}} + RX \longrightarrow R^{\bullet} + X^{-} + Ar$$

$$R^{\bullet} + e \rightleftharpoons R^{-}$$
$$R^{-} + H_2O \longrightarrow RH + OH^{-}$$

To be an effective catalyst the radical ion $Ar^{\bar{\cdot}}$ must be able to
exchange electrons with the organic substrate RX much faster
than the heterogeneous reaction by direct electron exchange
with the electrode. Electrogenerated radical ions with catalytic
abilities are usually aromatic species, and they are generated
at more negative potentials than the theoretical reduction potentials of the organic substrates but less negative than the
actual potential under the electrolysis conditions, as depicted
in the figure. The catalyst lowers the overpotential toward the
thermodynamic value of the RX compound (anodically electrogenerated catalysts should act analogously). Electrochemical transformation of secondary alcohols to their correspond-
ing chlorides has been homogeneously catalyzed by organic

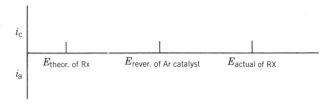

sulfides and phosphines, for example, Ph_2S and Ph_3P, in CH_2Cl_2 or $CHCl_3$ media. The catalytic species, Ph_2S^+, is produced anodically and forms sulfoxonium ion which reacts with Cl^- ion produced by the electroreduction of the chlorinated solvent.[47] In certain isolated cases redox chain catalytic mechanisms develop, for example, in the valence isomerization of quadricyclene.

Q

NBD

$$Q + e \xrightarrow[CH_2Cl_2]{} \overset{+\cdot}{Q} \longrightarrow N\overset{+\cdot}{B}D \xrightarrow{Q} NBD + \overset{+\cdot}{Q}$$

Because of the self-catalysis ($Q^{+\bullet}$ and $NBD^{+\bullet}$ are catalytic species) the current efficiency is much larger than the theoretical.[48] The very reversible quinone-hydroquinone couple has been studied as a mediator for both homogeneous and heterogeneous electrochemical reactions.[49] Electrogenerated adamanthylidine radical cations with oxygen catalyzed the synthesis of dioxetane via a chain reaction.[50] The oxidation of NADH to $NAD^{(+)}$ was catalyzed by chloranil acting by formation of a charge-transfer complex.[51] Also diimines[52] derived from diaminobenzenes were able to catalyze the oxidation of NADH. Electrogenerated anion of 1-ethyl-4-methoxycarbonyl pyridinium iodide was an active catalyst in the reduction of 1,2-dichloro-1,2-diphenylethane.[53] An important electrocatalytic reaction is the homogeneous dechlorination of the extremely toxic polychlorinated biphenyls, using 9,10-diphenylanthracene as a catalytic agent.[54] Organometallic complexes, especially those using N-macrocycles, for example, porphyrins, and phthalocyanines have been extensively studied, principally for the reduction of O_2 to H_2O for energy conversion purposes rather than organic synthesis applications. Electroreductive coupling of alkyl halides with Michael-type olefins is catalyzed by certain cobalt complexes, for example, vitamin B_{12} and similar model compounds, by forming alkyl-cobalt complexes as intermediates, which eventually lead to bicyclic ketones by 1,4 addition.[55] Mediated electron transfers via $Ru(bipy)_2Cl$ complexes are known to be effective for both homogeneous and heterogeneous catalysis, but no systematic applications to preparative electrosynthesis have been reported. Simpler organometallic complexes, for example, Mo(III) citrate, can act as effective redox catalysts. Thus CH_3CN in alkaline solution was reduced to give ethane, propane, and propylene, using a mercury cathode in the presence of molybdenum citrate.[56] Anodic substitutions of aryl mercury acetates are homogeneously electrocatalyzed by $Pd(OAc)_2$ in acetic acid media. Thus anisyl Hg(II) acetate afforded p-anisyl acetate in 98% yield, whereas no such reaction was evident without the catalyst.[57]

Redox catalysis by biphenyl, naphthalene, 2,6-dimethyl naphthalene, acenaphthene, and trimethyl phosphine was studied in the electrochemical reduction of fluorobenzene in DMF at mercury cathode. Catalytic currents were observed with all of these materials.[58]

Some indirect reductions can be assisted by a combination of a homogeneous catalyst and an inorganic ion; for example, the electrolytic reduction of thiophene was catalyzed by bi-

phenyl and zinc ion in DMF medium; however, chlorobenzene and methylcyclopropyl ketone were not catalyzed by such
a system.[59]

Recently an interest has been developing in preparing polymeric reagents in situ as redox catalysts by electrolyzing solutions containing a suitable monomer.[60]

3.18.4 Electrogenerated Bases

Certain organic molecules can be converted electrochemically
to very strong bases known as electrogenerated bases, EGBs.
Such bases are able to abstract protons from weak or very weak
organic acids. The substances from which the EGBs are generated are called *probases*. To be catalytic an EGB must be
capable of regeneration at the anode. EGBs can be used in
place of available bases which, although effective, may be too
dangerous to handle, or when no such bases are available. A
typical probase is azobenzene.

probase   EGB

Azobenzene can act as an EGB catalyst, as in the case below:

$$PhN=NPh \xrightarrow[-1.4V \text{ vs. sce}]{e} Ph\bar{N}-\overset{\bullet}{N}Ph$$

$$\downarrow RH \text{ (proton abstraction)}$$

$$PhNH\overset{\bullet}{N}Ph + R^-$$

$$PhNHNHPh \xrightarrow[[O]]{} PhN=NPh$$

$$\nearrow \text{PhNH}\overset{\bullet}{N}\text{Ph}$$

Thus using only catalytic amounts of azobenzene Michael-type condensations can be accomplished.[61]

$$CH_3CH_2NO_2 + \underset{\overset{|}{CH_3}}{{=\!\!\!\!/}C-CO_2CH_3}$$

$$\downarrow CH_3CN,\ EGB$$

$$CH_3\overset{\overset{O}{\parallel}}{\underset{}{C}}\diagdown\diagup\overset{CH_3}{\underset{}{\diagup}}\diagdown CO_2CH_3$$

Probases can be immobilized on electrodes so that the catalysis would be in effect heterogeneous, with its advantages and disadvantages, depending on specific cases.[62]

In order for an EGB to be of practical value it must be formed from a probase at as low as possible cathode potential, must be a relatively strong base, and, as mentioned before, it should be regenerated in situ in solution or at the anode. By controlling the potential, the rate of formation and concentration of the base can be controlled.[63] The overall electrochemical reaction rate can thus be regulated, and an otherwise difficult electroorganic reaction can be facilitated by the catalytic assistance of the EGB. The probase, for instance, $R_nC_6H_{5-n}$—N$=$N$C_6H_{5-n}R_n$, where R is a branched chain alkyl, C_3 to C_6, has been used to produce the EGB to deprotonate an acetic ester or derivatives thereof, $R'CH_2C(O)X$, where X can be OR″, NR″$_2$, halogen, and so on. Condensation of ethyl acetate was thus achieved at mercury pool cathode in DMF-Bu$_4$NBr medium.[64] Carbon tetrachloride in aprotic media is electroreduced to give a basic species which can assist certain reactions, for example, the cyclization.[65]

$$CCl_4 \xrightarrow[Q_4NBr/DMF]{-0.8\ V,\ sce} :\bar{C}Cl_3 + Cl^-$$

$$:\bar{C}Cl_3 + malonate(MeO_2C)_2CH(CH_2)_4Br \longrightarrow$$

$$\begin{array}{c} \text{CO}_2\text{CH}_3 \\ \dot{\text{CO}}_2\text{CH}_3 \end{array}$$

3.18.5 Superoxide Ion Reactions

Molecular oxygen can be electroreduced in aprotic media to superoxide ion:

$$O_2 \underset{-0.8\ V,\ sce}{\overset{e}{\rightleftharpoons}} O_2^{\cdot -}$$

The superoxide ion can abstract protons from acidic substances with $pK \leqq 0.5$ and thus can act as an electrogenerated base.[66] In the presence of H_2O it decomposes rapidly.[67]

$$O_2^{\cdot -} \longrightarrow \bar{O}-\dot{O} \xrightarrow{H_2O} HOO^{\cdot} + OH^-$$
$$\downarrow e/\text{cathode}$$
$$HOO^-$$
$$\downarrow H_2O$$
$$OH^- + H_2O_2$$

As an electrogenerated base, $O_2^{\cdot -}$ has assisted in converting enones to cycloketones in the presence of an auxiliary organic acid,[68] such as Ph_2CHCN or $MeCH(CO_2Et)_2$. Using electrogenerated tetraalkylammonium superoxide and oxygen, ethyl-cyanoacetate was converted to ethylglyoxalate and to oxomalonate.[69] Phosphonium salts were converted into ylides using electrogenerated bases, dianions, from the reduction of fluorenes in DMF. Thus the Wittig alkene synthesis was performed in the presence of phosphonium salts and aldehydes.[70] The reaction can be shown schematically.

$$B \xrightleftharpoons{2e} B^{2-}$$

$$R\diagdown\overset{+}{P}Ph_3 \xrightarrow{B^{2-}\to BH^-} R\diagdown PPh_3$$

$$\nearrow^{R'CHO}$$

$$RCH=CHR' \quad (cis \text{ and } trans)$$

B=

R = H, Br
X = CN, CO₂Et

3.18.6 Chemically Modified Electrodes as Electrocatalysts

No significant applications of chemically modified electrodes have yet been made to organic electrosynthesis. However, recent advances with regard to durability and reproducibility of this kind of electrodes promise some quite interesting applications of these fascinating electrocatalysts.[71] Some typical reductions of alkyl halides at iron-porphyrin modified electrodes have been shown in the cases of benzyl bromide and hexachloroethylbenzene. Carbon cathodes with tetra(p-aminophenyl) prophyrinato Fe(III) attached onto them decreased the overpotentials of these halides by as much as 1 V relative to the unmodified electrode. The active catalytic center was believed to be the Fe(II) and Fe(I) state of iron.[72] Asymmetric electroreductions of carbonyls and gem-dihalides have been carried out on electrodes coated with poly-L-valine.[73] Ascorbic acid oxidation was catalyzed by monolayers of catechols or amino-phenols attached on carbon electrodes,[74] and by vinyl-ferrocene films on graphite electrodes.[75] Citraconic acid and 4-methylcoumarin were asymmetrically reduced at a poly-L-valine coated graphite electrode, giving 25 and 43% optical yields of methylsuccinic acid and 3,4-dihydro-4-methylcoumarin in a phosphate buffered medium.[76] Electrodes coated with metallotetraphenyl porphyrins catalyzed the reduction of bromoalkyl compounds, for example, $PhCHBrCH_2Br$, $PhCHBrCHBrPh$, and $CH_2BrCHBrCH_2Br$. The catalysts were complexes of Co(II) or Cu(II) tetrakis(p-amino-phenyl) porphyrins. The rate of electron transfer was independent of the amount of catalyst attached on the electrode surface, and the reactions appeared to be inner sphere rather than outer sphere reactions.[77]

Graphite electrodes modified with catechols with a pyrene side chain, for example, 4- 2(1-pyrinyl)vinyl catechol, catalyzed the oxidation of the coenzyme dihydro-nicotinamide adenine dinucleotide (NADH). Direct oxidation of NADH requires very high overpotentials. However, surface modification of the anode with PSCH2, which is 4-2-(1-pyrenyl) vinylcatechol, greatly facilitates (lowers the overpotential) by an EC catalytic mechanism:[78]

$$\underset{\text{electrode}}{\underline{PSCH2}} \xrightarrow[-2H^+]{-2e} PSC \xrightarrow[H^+]{NADH} PSCH2 + NAD^+$$

the net reaction being

$$NADH \longrightarrow NAD^+ + 2e + H^+$$

Electrodes chemically modified with quinone-type mediators were designed, and they catalyzed the oxidation of NADH at potentials near the NADH/NAD$^+$ potential (E° ~310 mV vs NHE). Quinones are capable of both electron transfer and hydride transfer routes.[79] Graphite electrodes with α-cyclodextrin attached on their surface selectively catalyzed the reduction of o-nitrophenol in the presence of p-nitrophenol.[80] Chemically modified electrodes may offer some distinct advantages as regards stereochemical synthesis. For example, electrochemical asymmetric oxidation of phenyl cyclohexyl sulfide to sulfoxide was attained on a poly(L-valine)-coated platinum electrode.[81] Asymmetric reductions of citraconic and mesaconic acids were observed on optically active poly(amino acid)-coated electrodes.[82]

3.19 SOLID POLYMER ELECTROLYTE ELECTROLYSIS

Recently a new methodology is emerging in electrochemistry. New in the sense that it avoids the use of a supporting electrolyte in solution and replaces it by the so-called solid polymer electrolyte,[83] SPE. A special combination of anode-cathode arrangement is constructed with an ion exchange membrane *sandwiched* between the anode and cathode, as shown schematically below, the membrane assuming the role of the supporting electrolyte. The perfluorinated polymer membrane *Nafion* 415 has been mostly used. A number of organic syntheses have been carried out by the SPE method, for example, the Kolbe reaction and variations thereof, and electrocatalytic synthesis, as the indirectly catalyzed oxidation of cyclohexanol to cyclohexanone, using the I^-/I^+-cycle catalyst.[84] Both sides of the SPE membrane contained deposited platinum particles formed by hydrazine reduction in chloroplatinic acid solution. The practical advantage of this methodology is that it facilitates product separation and sometimes overcomes unwanted side reactions.

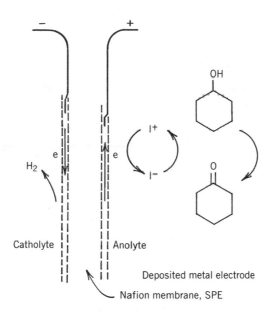

Incorporation of particles with catalytic activity for certain electrochemical reactions on one or both sides of the SPE membrane offers interesting electrocatalytic possibilities. Thus olefinic bonds have been hydrogenated by using SPE cation exchange membrane containing deposited electrocatalytic metals as electrodes.[85]

C. INDUSTRIAL STATUS OF ORGANIC ELECTROCHEMISTRY IN THE 1980S

As we overview the international chemical literature we meet thousands of research studies and patents dealing with organic electrochemistry. Yet, only a few large scale electroorganic processes are in operation worldwide. It seems that classical electroorganic synthesis cannot compete economically with conventional methods of synthesis on the industrial level; and it appears that the best hope for organic electrochemistry to ascend to industrial levels comparable to those of other methodologies is to become *electrocatalytic*. Indeed, several electroorganic processes that have come into existence have been more or less electrocatalytic, but no real efforts have been made to improve on and expand such processes. Unfamiliarity with the field on the part of older organic chemists and chemical engineers may be one reason for the lack of progress in electroorganic technology. Another reason may be the general belief that electrochemical processes are complex and difficult to scale up. However, a new interest has emerged recently worldwide in reexamining electrosynthetic methodology and applying it in various ways to the production of certain valuable products.

From what has been revealed in the open chemical literature, it seems that there are about one hundred electroorganic processes worldwide, ranging from small to medium scale of production. Most are spread over Germany, the United Kingdom, India, Russia, and Japan. The largest one is in the United States, which is Monsanto's adiponitrile process with an annual total production of 450 million pounds of adiponitrile, an intermediate for Nylon. Table 3.1 lists various commercial processes (some additional processes are included in Section B, Chapter 2). A much wider list is in Chemical and Engineering News, Nov. 19, 1984, p43, produced by R. Jansson, Monsanto, U. S. A.

It would be entirely outside the scope of this book to delve into the economics of electrochemical processes. Very recently efforts on a high level have been made by a few academic electrochemical engineers to develop engineering models and approaches suitable to modern electroorganic production works (C. Tobias, University of California; R. C. Alkire, University of Illinois; G. Sakellaropoulos, University of Thessaloniki). The research organic electrochemist must consider three basic criteria when planning for large-scale electrosynthesis: (1) Product selectivity, (2) electrical energy needs, and (3) capital requirements for the electrolysis system. The possibilities of using less expensive feedstocks than in chemical processes can be considered along with selectivity for the reaction. Electrocatalysis can save energy and also contribute greatly to the economy of the process by using only catalytic amounts of redox reagents. The possibility of in situ production of exotic reagents, or reagents that are too hazardous otherwise for use on any production scale, is an inherent special ability of electroorganic chemistry that has not yet been appreciated, despite the exhaustion of natural resources and the pollution of our environment.

TABLE 3.1 Some Commercial-Scale Electroorganic Processes

Process		Manufacturer
Acrylonitrile	$\longrightarrow$ Adiponitrile	Monsanto (USA; UK; Asahi, Japan)
Tetrahydrocarbazole	Hexahydrocarbazole	BASF, Germany
Phthalic anhydride	Dihydrophthalic Acid	BASF, Germany
Furan	2,5-Dimethoxy-2,5-dihydrofuran	BASF, Germany
4-*tert*-Butoxytoluene	4-*tert*-Butoxybenzaldehyde	BASF, Germany
2-Butyne-1,4-diol	Acetylene dicarboxylic acid	BASF, Germany
Monomethyl adipate	Dimethyl sebacate	BASF, Germany
p-Anisidine	Nitrobenzene	BASF, Germany
o-Aminobenzyl alcohol	Anthranilic acid	BASF, Germany
Anthracene	Anthraquinone	Holiday, UK
o-Nitrophenol	*o*-Aminophenol	India
3-Cyanopyridine	3-Aminomethylpyridine	India; USSR
Nitrobenzenes	Benzedines	India

o-Nitrotoluene	o-Toluidine	India
Lactose	Calcium Lactobionate	India (Sandoz)
Adiponitrile	1,6-Hexanediamine	USSR
Toluene-o-sulfonamide	Saccharin	USSR
Naphthalene	α-Naphthol	Union Carbide, US; Eastman-Kodak, US; BASF
Nitrourea	Semicarbazide	USSR
Phthalimide	Isoindole	Ciba-Geigy, USA
2-Methyl indole	2-Methyl indolene	Holiday, UK
Ethanol	Bromoform	India
Pyrene	Pyrenequinone	Holliday, UK
Salicylic Acid	Salicylaldehyde	USSR; India
Nitrourea	Semicarbazide	USSR

119

References

CHAPTER 3

Section A

3.1 and 3.2

1. Lund, H. and Iversen, P., in *Organic Electrochemistry*, Baizer, M. M., Ed., Marcel Dekker, New York, **1973**, p. 165; Muck, D. L. and Wilson, E. R., *J. Electrochem*. Soc., **117,** 1358 (1970).

2. Mirkind, L. A., Kornienko, A. G., Fioshin, M. Ya, and Bogdanova, N. P., *Electrokhimiya*, **15,** 413 (1979); Tyurin, Yu. M., Smirnova, L. A., and Naumov, V. I., *Elecktrokhimiya*, **15,** 445 (1979).

3. Lindsey, R. and Petersen, M., *J. Amer. Chem. Soc.*, **81,** 2073 (1959); Vasiliev, Yu. B., Bagotzky, V. S., and Korsman, E. P., *Electrokhimiya*, **27,** 929 (1982).

4. Einhor, J., Soulier, J. L., Bachuet, C., and Lelandais, D., *Can. J. Chem.*, **61,** 584 (1983).

5. Matsuda, Y., Kimura, K., Iwakura, C., and Tamura, H., *Bull. Chem. Soc.*, Japan, **46,** 430 (1973).

6. Weinberg, N. L. and Reddy, T. B., *J. Amer. Chem. Soc.*, **90,** 91 (1968); Hand, R. and Nelson, R. F., *J. Electrochem. Soc.*, **117,** 1353 (1970); Astad, B., Ronlan, A., and Parker, V. D., *Acta Chem. Scand.*, **B35,** 247 (1981).

7. Parker, V. D. and Ronlan, A., *J. Amer. Chem. Soc.*, **97,** 4714 (1975).

8. Becker, J. Y., Miller, L. L., Boekelheide, V., and Morgan J., *Tetrahedron Lett*, **1976,** 2939.

9. Parker, V. D. and Eberson, K., *J. Chem. Soc. Chem. Commun.*, **1972,** 441.

10. Callot, H. J., Louati, L., and Gross, M., *Bull. Soc. Chimique*, **II,** 317 (1983).

11. Baggaley, A. J. and Brettle, R., *J. Chem. Soc.*, **C**, 969 (1968).

12. Eberson, L. and Helgee, B., *Acta Chem. Scand.*, **B32**, 157 (1978).

13. Parker, V. D., *J. Chem. Soc. Chem. Commun.*, **1968**, 1164.

14. Scott, A., Dotson, P. A., McCapra, F., and Meyers, M. B., *J. Amer. Chem. Soc.*, **85**, 3702 (1963).

15. Ross, S. D., Finkelstein, M., and Petersen, R. C., *J. Amer. Chem. Soc.*, **86**, 4139 (1964); Scheele, J. J. and Eberson, L., *Acta Chem. Scand.*, **B32**, 36 (1978).

3.3

1. Adams, R. N., *Accounts Chem. Res.* **2**, 175 (1969).

2. Parker, V. D., *Acta Chim. Scand.*, **24**, 2757 (1970).

3. Nishiguchi, N. and Hirashima, T., *J. Org. Chem.*, **50**, 539 (1985).

4. Koch, D., Schäffer, H. and Steckman, E., *Chem. Ber.*, **107**, 3640 (1974).

5. Harrison, J. A. and Mayne, J. M., *Electrochim. Acta*, **28**, 1223 (1983).

6. Becker, J. Y. and Zinger, B., *J. Chem. Soc.*, Perkin (II), **1982**, 395.

7. Halter, M. A. and Malloy, T. P., U. S. Patent, 4212710, (1980); Kinoshita, T., Harada, J., Ito, S. and Sasaki, K., *Angew. Chem.*, **95**, 504 (1983).

8. Matsuda, Y., Nishiki, T. and Nakagawa, K., *Denki Kagaku*, (Japan), **52**, 199 (1984).

3.4

1. Sandholm, G., *Acta Chem. Scand.*, **25**, 3188 (1971); Finkelstein, M. and Ross, S. D., *Tetrahedron*, **1972**, 4497; Mayeda, E. A., Miller, L. L., and Wolf, J. F., *J. Amer. Chem. Soc.*,

94, 6812 (1972); Fleischmann, M., Korinet, K., and Pletcher, D., *J. Chem. Soc. Perkins Trans.*, **2,** 1396 (1972).

2. Adzic, R. R., Simic, D. N., Despic, A. R., and Drazic, D. M., *Z. Phys. Chem.*, N. F., **98,** 95 (1975); Vassiliev, Yu. B., Bagotzky, V. S., Korsman, V. A., Grinberg, L. S., Kavesky, L. S., and Polishchyak, V. R., *Electrochim. Acta*, **27,** 919 (1982); *Electrochim. Acta*, **27,** 929 (1982); Spasojevic, D. M., Adzic, R. R., and Despic, A. R., *J. Electroanal. Chem.*, **109,** 261 (1980); Jannakoudakis, A. D. and Kokkinidis, G., *J. Electroanal. Chem.*, **134,** 311, (1982); Kokkinidis, G. and Jannakoudakis, D., *J. Electroanal. Chem.*, **113,** 307, (1982).

3. Tabakovik, I., Grujic, Z., and Bejtouik, Z., *J. Heteroc. Chem.*, **20,** 635 (1983).

4. Remonov, B. S., Ponkratou, V. P., Abrutskay, I. A., and Fioshin, M. Ya., *Sov. Electrochem.*, **16,** 749 (1980).

5. Kaulen, J. and Schaeffer, H. J., *Tetrahedron*, **38,** 3299 (1982).

6. Shono, T., Matsumura, Y., Hayashi, J., and Usui, M., *Denki Kagatu Oyobi*, **51,** 131 (1983).

7. Zutshi, K., Rastogi, R., and Dixit, G., ISE, 34th Meeting, Erlagen, Germany, Sept. 1983.

8. Campbell, K. D., Gulbenkian, A. H., Edamura, F. Y., and Kyriacou, D., U. S. Patent 4,496,440 (1985).

9. Dixit, G., Rastogi, R., and Zutshi, K., *Electrochim. Acta*, **27,** 561 (1982).

10. Macagno, V. A., Vilche, J. R., and Arvia, A. J., *J. Electrochem. Soc.*, **129,** 301 (1982).

3.5

1. Masui, M. and Sayo, H., *J. Chem. Soc.*, **1968,** 973; *J. Chem. Soc.*, **1971,** 1593; Wawzonek, S. and McIntyre, T., *J. Electrochem. Soc.*, **114,** 1025 (1967); Hand, R. L. and Nelson, R. F., *J. Electrochem. Soc.*, **125,** 1059, (1978).

2. Andreades, S. and Zahnow, E., *J. Amer. Chem. Soc.*, **91**, 4181 (1969).

3. Matsuda, Y., Shono, A., Iwakura, C., and Ohshiro, Y., *Bull. Chem. Soc.* (Japan), **44**, 2960 (1971).

4. Feldhues, V. and Schafer, H. J., *Synthesis*, **82**, 145 (1982).

5. Cauquis, G. and Pierre, G., *C. R. Acad. Sci.*, **272**, 609 (1971); Yoshida, K. and Fueno, T., *J. Org. Chem.*, **37**, 4145 (1972).

6. Shono, T., Matsumura, Y., and Kashimura, S., *J. Org. Chem.*, **48** (19) 3388 (1983).

3.6

1. Eberson, L. and Nilsson, S. *Discuss. Faraday Soc.*, **45**, 242 (1968); Miller, L. L., *Pure Appl. Chem.*, **51**, 2125 (1979).

2. Nyberg, K., *Chem. Commun.*, **1969**, 774.

3. Laurent, E., Rauniyar, G., and Thomalla, M., *J. Appl. Electrochem.*, **14** (6), 741 (1984); *J. Appl. Electrochem.*, **15** (1) 1985.

4. Laurent, E., Rauniyar, G., and Thomalla, M., *Bull. Soc. Chim.*, (1-2, Pt. 1) 78 (1984).

5. Roult, E., Sarrazin, J., and Tallec, A., *J. Appl. Electrochem.*, **14** (5), 639 (1984).

6. Popp, G., *J. Org. Chem.*, **37**, 3058 (1972).

3.7

1. Watanabe, N., *J. Fluorine Chem.*, **22**, 205 (1983) (review).

2. Drakesmith and Hughes, F. G., *J. Appl. Electrochem.*, **9**, 685 (1979).

3. Laurent, E., Tardivel, R., and Thiebault, H., *Tetrahedron Lett.*, **24** (9), 903 (1983).

4. Simons, J. H., *J. Electrochem. Soc.*, **95**, 47 (1949).

3.8

1. Danly, D. E., *Chemistry and Industry*, **1979,** 439; Petrovich, J., Anderson, J. D., and Baizer, M. M., *J. Org. Chem.*, **31,** 3897 (1966); Rance, H. C. and Coulson, J. M., *Electrochim. Acta*, **14,** 283 (1969); Baizer, M. M., U. S. Patent 3,193,408 (1963).

2. Baizer, M. M., *Chemistry and Industry*, **1979,** 435.

3. Barger, H. J., *J. Org. Chem.*, **34,** 1489 (1969).

4. Degrand, C., Grodemaugo, C., and Compagnon, P. L., *Tetrahedron Lett.*, **1978,** 3023.

3.9

1. Popp, F. D. and Schultz, P. H., *Chem. Rev.*, **62,** 29 (1962); Fawcett, W. R. and Lasia, A., *Can. J. Chem.*, **59,** 3256 (1981); Sakamoto, M. and Takamura, K., *Bioelectrochem. Bioeneg.*, **9,** 571 (1982); Fournier, F., Berthetot, J., and Pascal, Y. L., *Can. J. Chem.*, **61,** 2121 (1983); Zuman, P., Barns, D., and Ryuolova, A., *Discus. Faraday Soc.*, **45,** 202 (1968).

2. Curphey, T. J., Amelloti, C. W., Layoff, T. P., MaCartney, R. I., and Williams, J. H., *J. Amer. Chem. Soc.*, **91,** 2817 (1969).

3. Degrand, C., Compagnon, P. L., and Belot, G., *Electrochim. Acta*, **29,** 5 (1984).

4. Shono, T., Kashimura, S., and Ishizaki, K., *Electrochim. Acta*, **29,** 605 (1984).

3.10

1. Wawzonek, S. and Fredicson, J. O., *J. Amer. Chem. Soc.*, **75,** 3985 (1955).

2. Sayo, H., Tsubitani, Y., and Masui, M., *Tetrahedron*, **1968,**

1717; Pearson, J., *Trans. Faraday, Soc.*, **44**, 683 (1948); Iversen, P. E. and Lund, H., *Acta Chem. Scand.*, **19**, 2303 (1965).

3. Markuez, J. and Pletcher, D., *Electrochim. Acta*, **26**, 17 (1981).

4. Shono, T., Hamaguchi, H., Mikani, H., Nogusa, H., and Kashimura, S., *J. Org. Chem.*, **48**, 2103 (1983).

5. Kokkinidis, G. and Jannakoudakis, P. D., *Electrochim. Acta*, **29**, 828 (1984).

6. Matsuda, Y., Kimura, H., and Okuhama, Y., *Denki Kagaku*, **52**, 796 (1984).

3.11

1. Donahue, J. and Oliver, J. W., *Anal. Chem.*, **41**, 753 (1969); Allen, M. J. and Steinman, K., *J. Amer. Chem. Soc.*, **74**, 3932 (1952); Fleszar, B. and Saneski, P., *Electrochim. Acta*, **27**, 429 (1982).

2. Kimura, M., Matsubara, S., and Sawaki, Y., *J. Chem. Soc. Chem. Commun.*, **1984**, 1619.

3. Saeva, F. D., Morgan, B. P., Fisher, M. W., and Haley, N. F., *J. Org. Chem.*, **49**, 390 (1984).

4. Yoshida, J., Safukuku, H., and Kawalata, N., *Bull. Chem. Soc.*, Japan, **56**, 1243 (1983).

3.12

1. Wawzonek, S. and Wagenknecht, J. H., *J. Electrochem. Soc.*, **110**, 420 (1963); Cipris, D., *J. Appl. Electrochem.*, **8**, 537 (1978); Osa, T., Fugihara, M., and Matsue, T., *J. Electrochem. Soc.*, **126**, 500 (1979); Matsuda, Y. and Hayashi, H., *Chemistry Lett. Chem. Soc.*, Japan, **1981**, 661; Takasu, Y., Matsuda, Y., Shimizu, A., Morita, M., and Saito, M., *Chemistry Lett. Chem. Soc.*, Japan, **1981**, 1685.

2. Barba, F., Guirado, A., and Zapata, A., *Electrochim. Acta*, **27**, 1335 (1982).

3. Wawzonek, S. and Wagenknecht, J. H., *J. Electrochem. Soc.*, **110**, 420 (1963); deLuca, C., Inesi, A., and Rampazzo, L., *J. Chem. Soc. Perkins Trans.* II, **1983**, 1821.

4. Rifi, M. R., *J. Amer. Chem. Soc.*, **89**, 4422 (1967).

5. Kyriacou, D., Edamura, F., and Love, J., U. S. Patent 4,217, 185 (1980).

6. Miller, L. L. and Hoffman, A. K., *J. Amer. Chem. Soc.*, **89**, 593 (1967); Beker, J. Y., *J. Org. Chem.*, **42**, 3997 (1977).

7. Abeywikrema, R. S., Della, E. W., and Fletcher, F., *Electrochim. Acta*, **27**, 343 (1982).

8. Matsuda, Y., Nishiki, T., Sakota, N., and Nakagawa, K., *Electrochim. Acta*, **29**, 35 (1984).

9. Takasu, Y., Matsuda, Y., Harada, M., and Masaki, M., *J. Electrochem. Soc.*, **13**, 350 (1984).

10. Matsuda, Y., Terashima, A., and Nakagwa, K., *Denki Kagaku*, **51**, 791 (1983).

3.13

1. Schiffrin, D. I., *Faraday Discuss. Chem. Soc.*, **56**, 75 (1974); Freese, K. W. and Canfield, D., *J. Electrochem. Soc.*, **131**, 2518 (1984); Kapusta, S. and Kackerman, N. J., *Electrochem. Soc.*, **131**, 1511 (1984); Silvestry, G., Gambino, S., Filardo, C., and Gulotta, A., *Angew. Chem.*, **96**, 978 (1984).

2. Zakaryan, A. V., Rotenberg, Z. A., Osetrova, N. V., and Vasil'eu, Y. B., *Sov. Electrochem.*, **14**, 1520 (1978).

3. Petrova, C. N. and Efimov, O. N., *Electrokhimiya*, **19**, 978 (1983).

4. Tezuka, M., Yajima, T., Tsuchia, A., Matsumoto, Y., Ushida, Y., and Hiday, M., *J. Amer. Chem. Soc.*, **104**, 6834 (1982).

5. Tyssee, D. A., U. S. Patent 3,945,896 (1976).

6. Ruttinger, H. H., Rudolf, W. D., and Matchines, H., *Electrochim. Acta*, **30** (2), 155 (1985).

7. Barba, F., Guirado, A., and Zapata, A., *Electrochim. Acta*, **27**, 1335 (1982).

8. Kitahara, S. and Osa, T., *Denki Kagaku Oyobi Kogyo*, **50** (9), 732 (1982).

9. Armand, J., Bellec, C., Boulares, L., and Pinson, J., *J. Org. Chem.*, **48**, 2847 (1983).

3.14

1. Nelson, R. F. in Weinberg, N. L. Ed., *Techniques of Electroorganic Synthesis*, Part II, Wiley-Interscience, New York; 1975, p. 269; Weinberg, N. L. and Weinberg, H. R., *Chem. Rev.*, **68**, 449 (1968); Dryhurst, G., *J. Electroanal. Chem.*, **28**, 33 (1970); Neptune, M. and McCreery, R. L., *J. Medic. Chem.*, **22**, 196 (1979); Bellamy, A. J. and Innes, D. C., *J. Chem. Soc. Perkin Trans. II*, **1982**, 1599.

2. Czpcjvalska, B. and Elving, P., *Electrochim. Acta*, **26**, 1755 (1981).

3. Macor, K. A. and Spiro, T. G., *J. Amer. Chem. Soc.*, **105** (7), 5601 (1983).

3.15

1. Nonaka, T. and Asai, M., *Bull. Chem. Soc.*, Japan, **51**, 2976 (1978).

2. Gottlieb, R., *J. Amer. Chem. Soc.*, **98**, 1708 (1976).

3. Kyriacou, D., Unpublished work.

4. Grimshaw, J., Hamilton, R., and Grimshaw, T., *J. Chem. Soc., Perkins Trans. I*, **229** (1982).

3.16

1. Tuck, D. G., *Pure and Appl. Chem.*, **51**, 2005 (1979); Nalco Chem. Co., U. S. Patent 3,256,161 (1961); Lehmkul, H., in M. M. Baizer, Ed., *Organic Electrochemistry*, Marcel Dekker, New York, 1973, p 621.

2. Jennings, P. W., Pillsbury, D. G., Hall, J. L., and Brice, V. T., *J. Org. Chem.*, **41**, 719 (1976).

3. Bustos, L., Green, J., Hencher, J. L., Khan, M. A., and Tuck, D. G., *Can. J. Chem.*, **61**, 2141 (1983); Kumar, N. and Tuck, D. J., *Can. J. Chem.*, **60**, 2579 (1982); Perichon, J., *Actual Chim.*, **9**, 25 (1982) (a review).

Section B

1. Torie, S., *J. Org. Chem.*, **46**, 3312 (1981).

2. Alkire, R. C. and Tsai, J. E. (University of Illinois).

3. Fleischmann, M., Korinek, K., and Pletcher, D., *J. Chem. Soc., Perkins Trans. (II)*, **1972**, 1396.

4. Fremery, M., Hover, H., and Schwarzlose, G., *Chem. Ing. Tech.*, **46**, 635 (1974).

5. Shaojun, D. and Kuwana, T., *J. Amer. Chem. Soc.*, **104**, 5514 (1982).

6. Laurent, E., Rauniyar, G., and Thomalla, M., *J. Appl. Electrochem.*, **14**, 741 (1984); **15**, 121 (1985).

7. Shono, T., Yoshihiro, Y., Hayashi, J., and Mitzoguguchi, M., *Tetrahedron Lett.*, **1979**, 165.

8. Rastogi, R., Dixit, G., and Zutshi, K., *Electrochim. Acta*, **29**, 1345 (1984).

9. Bizot, J. and Deprez, D., Ger(East) Patent DD210082 (1984).

10. Dixit, G., Rastogi, R., and Zutshi, K., *Electrochim. Acta*, **27**, 561 (1982).

11. Franklin, T. C. and Hond, T., *Electrochim. Acta*, **23**, 439 (1978).

12. White, D. and Coleman, J. P., *J. Electrochem. Soc.*, **125**, 1401 (1978); Jansson, R. E. W. and Tomov, N. R., *Electrochim. Acta*, **25**, 497 (1980).

13. Millington, J. P., *J. Chem. Soc.*, **B8**, 982 (1969); Evans, J. F. and Blount, H. N., *J. Org. Chem.*, **41**, 516 (1983).

14. Shono, T., Matsumura, Y., and Inoue, K., *J. Org. Chem.*, **48**, 1388 (1983).

15. Torii, S., Tanaka, H., and Ukita, M., Jap. Patent 5138872 (1979).

16. Zuerev, V. A. and Milman, V. I., *Electrokhimiya*, **16**, 1867 (1980).

17. Jap. Patent Kokkyo Koho 5839695 (1981).

18. Nakamura, Y., *Nippon Kagaku Kaishi, Chem. Abst.*, **98**, 34800 (1983).

19. Fioshin, M. Ya, Gromova, E. V., and Sorokouskh, A. T., *Electrokhimiya*, **17**, 1380 (1981).

20. Srinivasen, R. K. and Pathy, M. S. V., *J. Electrochem. Soc.*, India, **30**, 94 (1981).

21. Udupa, H., Indian Patent, 144,210 (1978).

22. Halter, M. and Mallroy, T. P., U. S. Patent, 4,212,711 (1980).

23. Jiang, L. C. and Pletcher, D., *J. Electroanal. Chem.*, **152**, 157 (1983).

24. Kinoshita, T., Harada, J., Ito, S., and Sasaki, K., *Angew. Chem.*, **95**, 504 (1983).

25. Anson, F. C., Collins, T. J., Gipson, S. L., and Kraft, T. E., *Amer. Chem. Soc. Abstracts of Papers*, 185th meeting, March, 1983.

26. Anson, F. C. Isou, Yu, Min, and Saveant, J. M., *J. Electroanal. Chem.*, **178**, 113 (1984).

27. Groves, J. T. and Murry, T. J., *Amer. Chem. Soc. Abstracts of Papers*, 186th Meeting, April, 1984.

28. Gilet, M., Mortreux, A., Folest, J. C., and Petit, F., *J. Amer. Chem. Soc.*, **105,** 3876 (1983).

29. Torii, S., Uneama, K., Ono, M., and Bannou, T., *J. Amer. Chem. Soc.*, **103,** 4606 (1981).

30. Torii, S., Okumoto, H., and Tanaka, H., *Chem. Letters*, Japan, **1980,** 617.

31. Thompson, M. S., DeGiovani, W. F., Moyer, B. A., and Meyer, T. J., *J. Org. Chem.*, **49,** 4972 (1984).

32. Jap. Patent, Kokai Tokyo Koho 5915792 (1983).

33. Pletcher, D. and Razak, M., *Electrochim. Acta*, **26,** 819 (1981).

34. Filando, G., LaRosa, F., Alfeo, G., Gambino, S., and Silvestri, G., *J. Appl. Electrochem.*, **13,** 403 (1983).

35. Kyriacou, D., U. S. Patent 4,242,183 (1980).

36. Giomini, C. and Inesi, A., *Electrochim. Acta*, **29,** 1107 (1984).

37. Brown, O. R., Middleton, P. H., and Threffal, T. L., *J. Chem. Soc. Perkins Trans. II*, **1984,** 955.

38. Miller, E. K., Swenson, K. E., Lehman, G. K., and Andruzz, R., *J. Org. Chem.*, **50,** 556 (1985).

39. Coleman, J. P. and Wagenknecht, J. H., *J. Electrochem. Soc.*, **128,** 322 (1981).

40. Murr, N. E., Tangi, J., and Payne, J. D., French Patent 2,542,764 (1983).

41. Udupa, H., Indian Patent 142241 (1979).

42. Noel, M., Ananthraraman, P. N., and Udupa, H., *Indian J. Techn.*, **19,** 100 (1981).

43. Rollin, Y., Meyer, G., Troupel, M., Fauvarque, J. F., and Perichon, J., *J. Chem. Soc. Chem. Commun.*, **15,** 793 (1983).

44. Rollin, Y., Meyer, G., Troupel, M., and Fauvarque, J. F., *Tetrahedron Lett.*, **1982,** 3573.

45. Fournier, F., Berthelot, J., and Pascal, Y. L., *Can. J. Chem.*, **61,** 2121 (1983).

46. Masui, M., Ueshima, T., and Ozaki, S., *J. Chem. Soc. Chem. Commun.*, **8,** 479 (1983).

47. Shono, T., Matsumura, Y., Hayashi, J., and Usui, M., *Denki Kagaku Oyobi Koggo*, **51,** 131 (1983).

48. Yasufuku, K., Takahashi, K., and Kutal, C., *Tetrahedron Lett.*, **25** (4), 4893 (1984).

49. Degrand, C. and Miller, L. L., *J. Electroanal. Chem.*, **117,** 267 (1981).

50. Clennan, E. L., Simmon, W., and Almgen, C. W., *J. Amer. Chem. Soc.*, **103** (8), 2098 (1981).

51. Huck, H. and Schmidt, H. L., *Angew. Chem.*, **93** (4), 421 (1981).

52. Kitani, A., So, Y. H., and Miller, L. L., *J. Amer. Chem. Soc.*, **103** (25), 7636 (1981).

53. Lund, H. and Kristensen, L. H., *Acta, Chem. Scand. Ser. B.*, **33** (7), 495 (1979).

54. Connors, T. I. and Rustling, J. F., *Amer. Chem. Soc.*, *Abstracts of Papers*, 186th Meeting, August, 1984.

55. Scheffold, R., Dike, M., Dik, S., Herold, T., and Walder, L., *J. Amer. Chem. Soc.*, **102** (10), 3642 (1980).

56. Petrova, G. N., Zueva, A. F., and Efimov, O. N., *Electrokhimiya*, **17** (6), 888 (1981).

57. Joensson, L., *Acta Chem. Scand.* Ser. B., **34** (9), 697 (1980).

58. Andrieux, G. P., Blockman, C., and Saveant, J. M., *J. Electroanal. Chem.*, **15** (2), 413 (1979).

59. Mairanovskii, S. G. and Kosychenko, L. I., *Soviet Electrochem.*, **16,** 266 (1980).

60. Yoshida, J., Sufuku, H., and Kawabata, N., *Bull. Chem. Soc.*, Japan, **56,** 1243 (1983); Chance, R. R., Boudeawx, D. S., and Bredas, J. L., *Amer. Chem. Soc. Abstracts of Papers*, 185th meeting, March, 1983.

61. Monte, W. T., Baizer, M. M., and Little, D. R., *J. Org. Chem.*, **48,** 803 (1983).

62. Degrand, G. and Laviron, E., *J. Electroanal. Chem.*, **117**, 283 (1981).

63. Saveant, J. M. and Binh, S. K., *J. Org. Chem.*, **42**, 1242 (1977).

64. Hallcher, R. C., Colden, R. D., and Baizer, M. M., U. S. Patent 4,293,393 (1980); Gooding, R. D., Hallcher, R. C., and Baizer, M. M., U. S. Patent 4248678 (1981).

65. Nugent, S. T., Baizer, M. M., and Little, R. D., *Tetrahedron Lett.*, **23** (13), 1339 (1982).

66. Sawyer, D. T. and Valentine, J. S., *Accounts Chem. Res.*, **14**, 393 (1981).

67. Jain, P. S. and Lal, S., *Electrochim. Acta*, **27**, 759 (1982).

68. Mitchio, S. and Baizer, M. M., *J. Org. Chem.*, **48**, 9931 (1983).

69. Sugawara, M. and Baizer, M. M., *Tetrahedron Lett.*, **24** (22), 2223 (1983).

70. Mehta, R. R., Pardini, V. L., and Utley, J. H. P., *J. Chem. Soc. Perkins. Trans. I*, **4**, 2921 (1982); Saveant, J. M. and Binh, S. K., *J. Org. Chem.*, **42**, 1242 (1977); Page, J. Y. and Simonet, J., *Electrochim. Acta*, **23**, 445 (1978); Hallcher, R. C., Goodin, R. D., and Baizer, M. M., U. S. Patent 4293393 (1981).

71. Faulkner, L. R., *Chem. and Eng. News*, Feb. 27, **1984**, p. 28.

72. Elliot, C. M. and Marree, C. A., *J. Electroanal. Chem.*, **119** (2), 359 (1981).

73. Abe, S., Fuchigani, T., and Nonaka, T., *Chem. Letters*, Japan, **7**, 1033 (1983); Abe, S. and Nonaka, T., Chem. Letters, Japan, **10**, 1541 (1983).

74. Stutts, K. J. and Wightman, R. M., *Anal. Chem.*, **55**, 1576 (1983).

75. Doutartas, M. F. and Evans, J. F., *J. Electroanal. Chem.*, **109**, 301 (1980).

76. Abe, S., Nonaka, T., and Fuchigami, T., *J. Amer. Chem. Soc.*, **105**, 3630 (1983).

77. Rocklin, R. D. and Murray, R. W., Report 1981, TR-15, Univ.

North Carolina, Gov. Rep. Accoun. Index (U.S.), **81** (5), 2360 (1981).

78. Jaegfeldt, H., Kuwana, T., and Johansson, G., *J. Amer. Chem. Soc.*, **105**, 1805 (1983).

79. Carlson, B. W. and Miller, L. L., *J. Amer. Chem. Soc.*, **107**, 479 (1985).

80. Tomokazu, M., Masamidi, F., and Totsuo, O., *J. Electrochem. Soc.*, **129** (8), 1681 (1982).

81. Komori, T. and Nonaka, T., *J. Amer. Chem. Soc.*, **105** (17), 5690 (1983).

82. Nonaka, T., Abe, S., and Fuchigami, T., *Bull. Chem. Soc.*, Japan, **56** (9), 2778 (1983).

83. Yeo, R. S., *J. Electrochem. Soc.*, **130**, 533 (1983); Lu, P. W. and Srinivasan, S., *J. Appl. Electrochem.*, **9**, 269 (1979); Oguni, Z., Yamashita, H., Nishio, K., Takahara, Z. I., and Yoshizawa, S., *Electrochim. Acta*, **28**, 1687 (1983).

84. Oguni, Z., Ohashi, S., and Takahara, Z., *Electrochim. Acta*, **30**, 121 (1985); Oguni, Z., Yamashita, H., Nishio, K., Takahara, Z., and Yoshizawa, S., *Electrochim. Acta*, **28**, 1687 (1983); Delue, N. R., U. S. Patent 4,472,252 (1984); Fugikawa, K. and Nakajima, H., *Electrochim. Acta*, **29**, 172 (1984).

85. Oguni, Z., Nishio, K., and Yoshizawa, S., *Electrochim. Acta*, **26**, 1779 (1981); Katayama, A. A. and Ohnishi, R., *J. Amer. Chem. Soc.*, **105**, 658 (1983).

Index

Acetone, reduction on various cathodes, 99

Activation, of electrodes, 21, 25, 32

Adsorption, 21, 22, 26, 58

Alcohols, carboxylic acids from, 99

Aldol condensation, electroinduced, 69

Alkanes, electrocatalytic oxidation, 97

Alkyl aromatics, indirect oxidation, 57

Alkyl halides, catalyzed reduction, 29

Alkyl iodides, as catalysts, 85

Alloy electrodes, 26

Amines, substitutions in, 63

Anisaldehyde, 97

Anodes, DSA, 32

Aromatics, in situ catalysts from, 37

Ascorbic acid, 113

Azobenzene, probase, 108

Bases, as catalysts, 56

Benzene, 93, 97

Benzofuroxan, 94

Benzyl alcohol, 61

Benzyl benzoate, 61

Birch-type reductions, 103

Butler–Volmer equation, 8

Carbon dioxide, 77
  and electrocarboxylations, 77–79

Carbon tetrachloride, basic species of, 110

Carboxylates, indirect oxidation, 15

Chemically modified electrodes, stereochemical synthesis, 114

Chiral electrodes, 30

137

Chloranil, catalyst in NADH oxidation, 107
Chlorophenoxyethanols, 60
Citraconic acid, 113
Colloidal particles, as electrodes, 33
Current density, mass transfer and primary products, 10, 11
Cyclic reactions, electroinitiated, 37
Cyclohexene, 49, 65

Desorption, 58
Diacetone-L-sorbase, anodic oxidation, 91
Dicarboxylations, 104
Dimethoxytetrahydropyrene, 50
Dimethyl malonate, 95
Diols, oxidation on NIOOH anodes, 59
Dioxetane, 107
Double layer, 2, 3
    hydrophobicity in, 14, 87
    and orientaion of species, 14
    and reaction course, 14, 67

Electric field, effects, 11, 101
Electroactive centers, 54
Electrocatalysis, 20, 28, 35, 37–39
    second kind, 87
    self-sustaining, 37, 107
Electrodes, 24, 26, 27, 29, 31–35
    adsorption on, 22, 26
    potential of, 5, 8, 10
Electrogenerated bases, requirements 110
Electrolysis medium, 12, 57, 64, 70
Electrolysis reaction, 2, 3
    basic steps, 2
    nature and products, 5, 7, 12
    theoretical potential for, 12
Electron hopping, 31
Electropolymerizations, 31
Electrosorption, 22
Electrosynthesis:
    large scale, 86–92
    paired, 91–95
    with SPE, 114
    table, 118

Feeder electrode, 33, 34
Fe(II)-Fe(I) catalytic centers, 113
Fe-phthalocyanine, redox centers, 31
Fluorinations, anodic, 66
Fluorobenzene, calatyic currents, 108

gem-dihalides, 113
Glassy carbon, oxide particles on, 93
Graphite electrode, catechols on, 113
Grignard reagents, electrosynthesis with, 85

Halide ions:
    catalysis with, 89, 93, 96
    substitutions, aromatics, 63
Halopyridines, 106
Hexadecane, 85
Hydrodimerizations, 67
Hydrogenations, 100

Intercalation, electrodes by, 35
Intermediates, active, 10
Iodo compounds, cyclization with, 83
Ion-pair catalysis, 75
Iodonium ion, electrocatalyst, 93–96, 115
Isobutanol, 58

Kolbe synthesis, 47
    and biomass, 50
    diacids by, 50
    with SPE, 114

Lead dioxide anode, in H_2SO_4 emulsions, 93
Lifetimes, of primary species, 12

Metallic complexes, formed in situ, 85
    catalysis by, 96, 97, 104, 105
Metallocarbenes, formation in situ, for olefin metathesis, 98
Methoxylations with SPE, 65
Methyl chloroformate, carboxylations, 78
Modified electrodes, as electrocatalysts, 112

Nation-coated electrodes, 31, 97
NiOOH, anodes, 59, 60, 92
N_4-macroycles, on electrodes, 29, 108
Nickel salts, catalysts, 58

Organic electrocatalysts, 106, 107
$Os(bpy)_3^{2+}$, 97
Overpotential, 9, 10, 25
Oxidants, regeneration, 15
Oxide electrodes, 32
Oxyanions, metallic, catalysts, 15

Periodate, catalytic oxidations, 89
Phase-transfer electrocatalysis, 39
Ph_2S and Ph_3P, catalysts species, 107
Pinacolization, catalysis in, 104
Poly-L-valine, films, 108
Polymeric reagents, catalysts, 72
Poly(pyrrole) films, 31
Porphyrin electrodes, 29, 113
Potential cycling, 32
Powder electrodes, 34

Quaternary salts, 39, 67, 86
Quinone-hydroquinone, as mediator, 107

Reaction rates, 8, 11, 14
Redox catalytic couples, 30, 33, 36

Reversibility and irreversibility, 7, 8
$Ru(bipy)_2Cl$, 33
$Ru(IV)$, 99

Saccarin, electrosynthesis of, 97
Selenenylating reagents as
    electrocatalysts, 98
Semiconductors, catalytic electrodes, 34
Silane reagents, 32
Specific catalysts, 11
Spongy silver electrode, 87, 88, 101
Substitution:
    in micellar media, 65
    by phase-transfer catalysis, 64
Sulfur compounds, 72
Superoxide ion, 111

Tafel equation, 8, 13
Thiolates, deprotection, 72
Transition metals, catalysts, 29, 57
Triphenylamine, catalyst, 15

Underpotential deposition, 27

Vitamin B_{12} and models of, as catalysts, 108

Water emulsions, oxidations in, 57
Wittig alkene synthesis, 112